Stem Cells: A Very Short Introduction

VERY SHORT INTRODUCTIONS are for anyone wanting a stimulating and accessible way in to a new subject. They are written by experts, and have been published in more than 25 languages worldwide.

The series began in 1995, and now represents a wide variety of topics in history, philosophy, religion, science, and the humanities. The VSI Library now contains 300 volumes—a Very Short Introduction to everything from ancient Egypt and Indian philosophy to conceptual art and cosmology—and will continue to grow in a variety of disciplines.

Very Short Introductions available now:

ADVERTISING Winston Fletcher
AFRICAN HISTORY
    John Parker and Richard Rathbone
AGNOSTICISM Robin Le Poidevin
AMERICAN IMMIGRATION
    David A. Gerber
AMERICAN POLITICAL PARTIES
    AND ELECTIONS L. Sandy Maisel
THE AMERICAN PRESIDENCY
    Charles O. Jones
ANARCHISM Colin Ward
ANCIENT EGYPT Ian Shaw
ANCIENT GREECE Paul Cartledge
ANCIENT PHILOSOPHY Julia Annas
ANCIENT WARFARE
    Harry Sidebottom
ANGELS David Albert Jones
ANGLICANISM Mark Chapman
THE ANGLO-SAXON AGE John Blair
THE ANIMAL KINGDOM
    Peter Holland
ANIMAL RIGHTS David DeGrazia
ANTISEMITISM Steven Beller
THE APOCRYPHAL GOSPELS
    Paul Foster
ARCHAEOLOGY Paul Bahn
ARCHITECTURE Andrew Ballantyne
ARISTOCRACY William Doyle
ARISTOTLE Jonathan Barnes
ART HISTORY Dana Arnold
ART THEORY Cynthia Freeland
ATHEISM Julian Baggini
AUGUSTINE Henry Chadwick
AUTISM Uta Frith
THE AZTECS David Carrasco

BARTHES Jonathan Culler
BEAUTY Roger Scruton
BESTSELLERS John Sutherland
THE BIBLE John Riches
BIBLICAL ARCHAEOLOGY
    Eric H. Cline
BIOGRAPHY Hermione Lee
THE BLUES Elijah Wald
THE BOOK OF MORMON
    Terryl Givens
THE BRAIN Michael O'Shea
BRITISH POLITICS Anthony Wright
BUDDHA Michael Carrithers
BUDDHISM Damien Keown
BUDDHIST ETHICS Damien Keown
CANCER Nicholas James
CAPITALISM James Fulcher
CATHOLICISM Gerald O'Collins
THE CELL
    Terence Allen and Graham Cowling
THE CELTS Barry Cunliffe
CHAOS Leonard Smith
CHILDREN'S LITERATURE
    Kimberley Reynolds
CHINESE LITERATURE Sabina Knight
CHOICE THEORY Michael Allingham
CHRISTIAN ART Beth Williamson
CHRISTIAN ETHICS D. Stephen Long
CHRISTIANITY Linda Woodhead
CITIZENSHIP Richard Bellamy
CLASSICAL MYTHOLOGY
    Helen Morales
CLASSICS
    Mary Beard and John Henderson
CLAUSEWITZ Michael Howard

Available soon:

For more information visit our website
www.oup.com/vsi/

Jonathan Slack

STEM CELLS
A Very Short Introduction

OXFORD
UNIVERSITY PRESS

Great Clarendon Street, Oxford ox2 6DP

Oxford University Press is a department of the University of Oxford.
It furthers the University's objective of excellence in research, scholarship,
and education by publishing worldwide in

Oxford New York

Auckland Cape Town Dar es Salaam Hong Kong Karachi
Kuala Lumpur Madrid Melbourne Mexico City Nairobi
New Delhi Shanghai Taipei Toronto

With offices in

Argentina Austria Brazil Chile Czech Republic France Greece
Guatemala Hungary Italy Japan Poland Portugal Singapore
South Korea Switzerland Thailand Turkey Ukraine Vietnam

Oxford is a registered trade mark of Oxford University Press
in the UK and in certain other countries

Published in the United States
by Oxford University Press Inc., New York

British Library Cataloguing in Publication Data

Data available

Library of Congress Cataloging in Publication Data

Data available

Typeset by SPI Publisher Services, Pondicherry, India
Printed in Great Britain
on acid-free paper by
Ashford Colour Press Ltd, Gosport, Hampshire

ISBN 978-0-19-960338-1

1 3 5 7 9 10 8 6 4 2

Contents

List of illustrations

The publisher and the author apologize for any errors or ommission in the above list. If contacted they will be pleased to rectify these at the earliest opportunity.

Preface

This book is intended to introduce readers to stem cells: to explain what they are, what scientists do with them, what therapy is available today, and what might be reasonably expected to happen in the next few years. It deals with the science and medicine of stem cells rather than with ethics, law, or politics. There are many other books, and endless media coverage, about these other issues, but this book focuses on the science.

I hope readers will be able to take away an understanding of the difference between embryonic stem cells and tissue-specific stem cells, and of the importance and potential of the recently discovered induced pluripotent stem cells. They should also learn that present stem cell therapy is still in its infancy. Most real stem cell therapy today is some type of bone marrow transplantation, and most of the other activities going under the name (called 'aspirational stem cell therapy' in this book) have limited rationale and are probably ineffective.

The application of most of the science still lies in the future. Accordingly, I write as an optimist and I believe that future generations will reap huge rewards from current stem cell research and from regenerative medicine research more generally.

I am grateful to the following for reading drafts of the book and advising on accuracy and accessibility: Janet Slack, Rebecca McKnight, Pamela Self, Helen Brittan, and Jakub Tolar.

Box 1: Essential Acronyms

In this book, abbreviations are kept to a minimum but a few are unavoidable:

ES cell: Embryonic stem cell
iPS cell: induced pluripotent stem cell
DNA: Deoxyribonucleic acid
FDA: Food and Drug Administration
HLA: Human leukocyte antigen
HSC: Haematopoietic (blood forming) stem cell
HSCT: Haematopoietic stem cell therapy

The Glossary at the end provides further explanation of these acronyms as well as definitions for all the terms introduced in *italic* text.

Chapter 1
What are stem cells?

It is an unpalatable fact that we all get old and we all die. We spend most of our lives trying not to think about this fact, and to help us in this regard most religions have created beliefs about survival of the self-conscious mind even after the body and brain have decayed. But while we tend to try not to think about death, all of us are very keen to avoid disability. We have a horror of congenital conditions such as cerebral palsy; of accidents that can cause serious injury such as blindness or paralysis; and of the loss of independence caused by many of the diseases that accompany old age, such as Alzheimer's disease, strokes, heart failure or cancer. This is especially the case if we have a close friend or family member who suffers from one of these conditions, or who cares for a sufferer. We yearn for some miracle cure that will end the suffering and restore the person we have known.

Our yearning is all the greater because of the success of public health in preventing, and of scientific medicine in curing, a large range of other afflictions of childhood and middle life. Most of us who live in rich countries can now expect to live to old age with only very mild medical problems on the way, and this makes all the more tragic those cases of serious disease that still do occur.

Most people value stem cell research because they believe it will generate new and effective cures for currently incurable conditions.

This is the main force behind the large sums of research money currently devoted to the area. Some scientists are also motivated by the desire to develop new cures, but often they have more limited objectives involving the understanding of particular biological phenomena. This is just one example of the different perceptions that different people have of stem cell research. When we consider that stem cell biology is also of interest to bioengineers, politicians, biotech investors, and desperate patients, and involves a bevy of ethical, legal and religious interests, we can see that this is a scientific area of more than usual fascination.

What is a stem cell?

A *stem cell* is a cell that can both reproduce itself and generate offspring of different functional cell types. To understand what this means we shall start by considering the nature of cells. Cells are the ultimate structural unit of an animal or plant body. Each has a *nucleus* containing the genetic material (*DNA*), and a *cytoplasm* containing a complex mixture of *proteins* and other sorts of molecule that perform particular biochemical or mechanical tasks. There are about 210 kinds of cell in a human body. Most of them are what we call *differentiated* cells, each type of which has a specific function and a particular appearance when viewed down the microscope. Cells of the liver (*hepatocytes*), of the heart muscle (*cardiomyocytes*), and of the brain (*neurons*), are well known types of differentiated cell. The differentiated type that a cell belongs to depends on which particular *genes* are active in its nucleus. Each gene encodes one specific protein and the repertoire of genes that are active, and thus of proteins that are produced defines the type of the cell. The term *gene expression* is used to describe the production of proteins from active genes. The complete set of genes present in the cell nucleus is called the *genome*, and to a first approximation the genome is the same for every cell in the body. An undifferentiated cell is one that does not have any obvious specialization of *gene expression* and has a bland generic appearance down the microscope. But just because you cannot see specialization this does not mean that it does not exist. Most undifferentiated cells are specialized in some way, especially in terms of restrictions into what other types of cell they can become. Undifferentiated cells are found in the *embryo*, where they develop into various types of differentiated cells in the course of time. They are also found in some cancers, where they tend to be bad news because of their capacity for unrestricted growth. Undifferentiated cells are sometimes, but by no means always, stem cells.

There is a fairly good consensus on the definition of stem cells given above. This comprises just two properties: they are cells that can reproduce themselves, and they can also generate daughter cells that become differentiated cells (Figure 1). Examples of differentiated cells arising from stem cells are those of the skin, the blood, and the lining of the intestine. Consider the skin as a familiar example. Its outer layer is called the epidermis and is composed of cells called keratinocytes. The top layer of cells of the epidermis are worn away every day, and our skin persists as a functional tissue because new cells are being created continuously in the lowest layer of cells. This basal layer contains the stem cells of the epidermis. When they divide, about 50 per cent of their progeny remain in the basal layer as stem cells and the others divide a few more times and then enter a programme of maturation to become keratinocytes. As they mature they move up through the layers of cells that comprise the epidermis. They start to activate new genes and to make proteins including large amounts of fibrous proteins called keratins which give the skin its desirable properties of flexibility, strength, and impermeability. Eventually the epidermal cells die and become flat discs largely composed of keratin. It is these dead cells that are constantly being rubbed away from the exterior of our skin.

The epidermis is an example of a *renewal tissue*, one whose cells are continuously being renewed by cell division throughout the life of the organism. Renewal tissues could not exist without stem cells, and the best characterized types of stem cell are those responsible

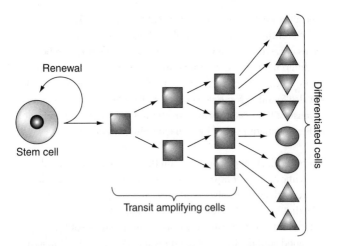

1. **The concept of a stem cell. The stem cell renews itself and generates differentiated cell progeny. The immediate progeny of a typical tissue-specific stem cell will be transit amplifying, or progenitor, cells that divide a finite number of times before differentiating. Often, but not always, the stem cell generates more than one type of differentiated cell**

for tissue renewal. They are often called *tissue-specific stem cells* to reflect the fact that each type of stem cell is responsible for making the cells of its own particular tissue and no other.

The structure of a renewal tissue always contains some form of microenvironment—called a *niche*—that is favourable for the persistence and function of the stem cells. For example, epidermal stem cells are associated with projections (papillae) from the base layer (dermis) in the skin; intestinal stem cells are associated with granule-containing Paneth cells at the base of tiny pits within the intestine (know as crypts); and blood-forming (*haematopoietic*) stem cells are located in the bone marrow, associated with bone cells or blood vessels.

Stem cells are by no means the only dividing cells in the body. Embryos, young organisms, and renewal tissues of adults all contain many other dividing cells that do not persist indefinitely and become some other cell type after a few cell divisions. These are called *progenitor* cells, or specifically in renewal tissues, *transit amplifying cells*.

The properties of the renewal tissues enabled the original definition of stem cell behaviour in terms of the ability to self-renew and to generate differentiated progeny. But the most famous stem cell of them all is now the *embryonic stem cell (ES cell)*. In one sense, the ES cell is the iconic stem cell. It is the type of stem cell that has attracted all of the ethical controversy, and it is what lay people are thinking of when they refer to 'stem cell research'. But ironically, the embryonic stem cell does not exist in nature. It is a creature that has been created by mankind and exists only in the world of *tissue culture*: the growth of cells in flasks in the laboratory, kept in temperature-controlled incubators, exposed to controlled concentrations of oxygen and carbon dioxide, and nourished by complex artificial media. Cells grown in culture are often referred to by the Latin phrase *in vitro* (in glass, since the relevant containers used to be made of glass) and distinguished from *in vivo* (inside the living body).

ES cells do satisfy the basic definition offered above: they are undifferentiated cells that can divide without limit; and they can also produce functional differentiated cells, probably all the cell types that are normally found in the body. ES cells originate from cells that lie within the early embryo. The reason that their *in vivo* counterparts are not regarded as true stem cells is that in normal embryonic development they will soon develop into other cell types, so, unlike the basal layer epidermal stem cells mentioned above, they do not remain the same for more than a few days. But *in vitro*, in the culture flask, they really are stem cells because they can either remain the same for years, or can be caused to differentiate to a range of functional cell types. To avoid confusion it is useful always to distinguish clearly between the tissue-specific stem cells, such as those in the epidermis, and the *pluripotent* stem cells, which comprise the embryonic stem cells and also the *induced pluripotent stem cells (iPS cells)* which closely resemble them.

The term *pluripotent* will be used a lot in this book. It means the ability to form any of the cell types found in the normal body. Tissue-specific stem cells are not pluripotent as they are able to form only the cell types of one tissue. They are sometimes described as multipotent when the tissue contains many cell types (like the blood), or unipotent if there is only one (like the sperm-forming stem cells of the testis).

Another term that will frequently be encountered on websites and in ethical debates is *adult stem cell*. This is not so much a biological term as a political one. It refers to anything that might be considered a stem cell but is not an embryonic stem cell. So tissue-specific stem cells and iPS cells are both counted as adult stem cells although they are completely different in their properties, and in fact iPS cells closely resemble ES cells. In addition, various ill-defined cells in culture are referred to as 'adult stem cells' even though they may originate from parts of the placenta, the umbilical cord blood, or juveniles, as well as from adult human beings.

Philosophically inclined readers may have noticed that the definition of the stem cell I have introduced, which is generally accepted among biomedical scientists, involves defining a behaviour rather than an intrinsic state. In other words we cannot identify a stem cell as being a particular sort of 'thing', we can only identify it by observing what it does. This is a real practical issue and not just a philosophical one. Many scientists have attempted to find genes whose expression is characteristic of all types of stem cell, but so far this endeavour has proved unsuccessful, except in the trivial sense that the genes required for cell survival or cell division are necessarily active in stem cells. In particular, the genes known to be responsible for the pluripotent behaviour of

ES cells are not normally active in tissue-specific stem cells. So we are stuck with the fact that we can only define stem cells by their behaviour, and that the best known type of stem cell, the embryonic stem cell, is a human-created artefact rather than an entity belonging to nature.

Tissue culture

As implied above, the technology of tissue culture, also called cell culture, is absolutely central to stem cell biology and so its characteristics need to be introduced. Tissue culture means growing cells outside the body, in tubes, dishes, or flasks filled with artificial media.

The first crude methods for keeping animal cells alive *in vitro* were introduced by embryologists in the late nineteenth century and were refined particularly by Alexis Carrel, a French scientist who worked in the early twentieth century at the Rockefeller Institute for Medical Research in New York (later Rockefeller University). But tissue culture only became widely practiced from the 1950s, when it became possible to purchase the required complex media from biological suppliers, and it became easy to suppress microbial contamination using the newly available antibiotics.

Cells within the body are mostly not dividing and they exist within tissues, usually as communities of several different cell types in close proximity to each other and to blood vessels and nerves. When small pieces of tissue are placed into a tissue culture environment, certain cells will migrate out of the pieces and start to grow, but many others will not, so tissue culture is intrinsically a selective process. Tissue culture cells are normally grown on plastic surfaces where they increase in numbers to form a single layer (monolayer). Once the monolayer is continuous and fills up the whole dish, the cells will usually stop growing. In order to keep them growing they are subcultured by treating with an enzyme

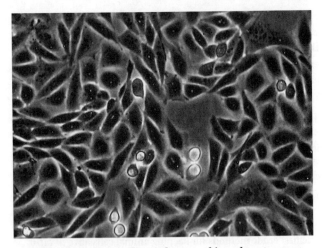

2. Cells growing in tissue culture. These are chinese hamster ovary cells. Each dark shape is one cell

(trypsin) which degrades their attachments to the plastic and allows them to rise into the medium. As a suspension, the cells can be diluted into fresh medium and dispensed into fresh containers where they will reattach and the growth cycle can begin again. Each such subculture is called a *passage*. All manipulations of the cells need to be carried out in special safety cabinets which have a sterile air supply to prevent any bacteria or fungal spores from entering the vessels. Because the cell culture media are highly nutritive, microorganisms will thrive on them and if they manage to enter the culture they will outgrow the animal cells very easily. The culture vessels are usually kept in incubators at 37°C (mammalian body temperature) and 5 per cent carbon dioxide (a level approximating the normal tissue environment).

Tissue culture cells can also be stored frozen in liquid nitrogen (–196°C). Using careful freezing and thawing techniques, they can be stored at this temperature indefinitely and later revived for continued growth.

Certain types of cell thrive in tissue culture and others do not. Most tissue pieces (explants) placed in culture will generate a growth of cells called *fibroblasts*, which have an irregular stellate shape and are oriented at random. These are thought to be similar to the cells normally found in the dermis (the lower layer of the skin), but the origin of tissue culture fibroblasts is not really known, nor is it known how many different types of cell in the body the fibroblasts that grow in culture actually represent. It is also possible to culture the other main morphological class of cell, called *epithelial cells*. Epithelial cells, of which the epidermal keratinocyte is one example, form sheets with attachments between the cells and normally secrete an extracellular layer, the basement membrane, beneath the sheet. The upper and lower surfaces of epithelial are different (the cells are said to be polarized) and in the epithelial sheet they are all oriented the same way up. Most of our functional tissues are composed of epithelia, for example the epidermis of the skin; the linings of the gut, reproductive, and respiratory tracts; the liver; and all of our glandular organs.

Tissue culture has been very useful for all sorts of purpose, especially growing cells under conditions where they are amenable for experimentation, and also for practical purposes such as the industrial production of vaccines or the various hormones and *growth factors* that are now used to some extent in therapy. However tissue culture has two important characteristics that can cause problems and it is especially important to understand these in the context of stem cell research. First, the tissue culture environment, being one of rapid growth, is highly conducive to selection. If a mutation occurs in one cell which gives it a small advantage in growth rate over all the others, then the progeny of that cell will soon overgrow the rest of the culture and come to represent virtually all the cells in the dish. This means that all tissue culture is an exercise in accelerated evolution and it is inevitable that after a certain number of passages the cells you get out are not quite the same as the ones you put in. Secondly, even without mutation and selection, cells can change their properties in tissue culture. Cells always need to be responsive to their environment, and the nutrients, growth factors, hormones and physical substratum experienced in tissue culture may alter the gene expression of cells quite profoundly and mean that they acquire markedly different properties from those which they had *in vivo*. As mentioned above, embryonic stem cells themselves do not quite correspond to any normal cell type in the embryo, and the pluripotent cells sometimes isolated from adult organisms are also widely believed to be artefacts of tissue culture.

The question of tissue culture artefacts raises for the first, but by no means the last, time in this book the very important issue of the differences of perception and outlook experienced by various kinds of professional engaged in this line of work. To a biological scientist, who wants to understand nature, if a cell population in culture has changed since it was in the body, then it is an artefact and it is intrinsically less interesting than its counterpart within the body. But to a bioengineer who wants to make useful products, the interest of the cells lies in their potential utility and their abnormal features may even be an advantage for the purpose envisaged. In other words, to the engineer the interest lies in what the cells can be made to do, and their place or otherwise in the natural world is of secondary importance. Members of the public are much more likely to identify with the world view of the engineer than that of the scientist. But most stem cell research has been done by scientists, and so their outlook tends to dominate the agenda and can sometimes lead to priorities that seem perverse to the general public.

Cell therapy

The reason that stem cell research is seen as the source for new cures is because this technology offers the possibility of cell therapy. This means making new cells to replace ones that have died, and introducing them into the appropriate part of the body in a functional state; for example, heart muscle cells into a failing heart, or neurons into a part of the brain affected by a stroke. If old, damaged or dying cells can be replaced by young, growing and functional cells, then organ damage could be reversed and the recipient could be cured of his or her disease and live for a lot longer. Stem cells are seen as the key to cell therapy because they are the source of the new cells for transplantation. The tissue-specific stem cells are the ones that normally produce many of the differentiated cell types of the body, and the embryonic stem cells can probably produce all cell types under suitable circumstances.

This sounds very attractive, and the websites of numerous companies and clinics offering 'stem cell therapy' will present matters in a simple way like this. But the reality of cell therapy presents more of a problem. The first difficulty is that many of the important cell types in the body do not grow during adult life. Once they have been formed in embryonic development, they tend to persist for long periods, perhaps even for life, with minimal cell turnover. But to generate enough cells for transplantation a useful cell source has to be expandable. In other words a few cells need to be able to multiply in tissue culture to become a large population suitable for grafting. Most of the useful cell types required for cell therapy do not divide at all in tissue culture, or if they do divide they rapidly lose the characteristics and properties that make them useful. This is true for example of liver cells (hepatocytes) and the insulin-producing cells of the pancreas (*beta cells*). Because of this problem, very little of the successful cell therapy carried out today involves expansion of cell populations in tissue culture prior to grafting. Current cell therapy mostly consists of grafting of cells directly from one person to another. One reason that bone marrow transplantation is the most important form of current stem cell therapy is that it is possible to harvest bone marrow (and the *haematopoietic stem cells* it contains) in sufficient quantity from living donors. Other types of cell therapy, such as grafts of pancreatic islets or hepatocytes, are generally done using human cadaver donors, which greatly limits the cell supply and the utility of the methods.

The next major problem confronting cell therapy is that of immunity. Human beings, and all other vertebrate animals, have a very complex immune system that evolved as a means for combating infection; cells of invading microorganisms are recognized as foreign and destroyed. This system also causes the rejection of cell or tissue grafts from one individual to another. The cells of the graft are recognized as 'not self' by a population of white blood cells called *T lymphocytes*. These destroy target cells by exposing them to toxic substances called *cytokines*, among which are the interferons, also important in the response to virus infection, and tumour necrosis factors, which are also sometimes secreted by tumours and account for many of their systemic toxic effects. The T lymphocytes can recognize almost any novel molecule that they do not normally encounter in the body, but the bulk of the immune response is directed against a family of cell surface molecules called *HLA (human leukocyte antigen)* factors. These are glycoproteins, molecules composed partly of protein encoded by the HLA genes and partly of carbohydrate. The HLA genes show enormous variability between individuals.

At this point it is worth mentioning that when biomedical scientists speak of a 'gene', they are usually thinking of the normal version of the gene, which encodes a specific protein having a specific function. But evolutionary biologists, or psychologists, or lay people, speak of a 'gene' they are usually thinking of the variants of a gene that different people may have, such that some of the difference of appearance, personality or behaviour is due to the variant possessed. Biologists call the variants of genes possessed by different individuals *alleles*. The HLA system comprises two clusters of genes, called ABC and DR, each of which may be occupied by any of a variety of alleles. The total number of possible combinations of these alleles is very large, which is why grafting of tissue from one individual to another (an *allogeneic graft* or *allograft*) usually provokes a rejection reaction by the T lymphocytes of the host. In clinical practice these reactions are controlled by drugs that suppress the immune system (*immunosuppressive drugs*). But the difficulty of doing this is increased if there is not too much mismatch between the HLA alleles of graft and host. Determining the degree of HLA match is the basis of tissue typing, which is very important both for organ transplantation and for the currently practiced types of cell therapy.

It has been known since the 1950s that tissue grafts are tolerated if they are carried out between identical twins. This is because identical twins arise by spontaneous division of the early embryo into two parts. They have exactly the same set of alleles for all of the HLA genes, and for any other genes involved with graft rejection. The same is true of any graft taken from an individual and reimplanted back into another position in the same individual (an *autologous graft*, or *autograft*).

The use of immunosuppressive drugs to control the rejection of allografts comes at considerable cost. The drugs often have side effects causing various types of organ damage and, because they reduce the ability of the individual to mount an immune reaction against invading microorganisms, they also make immunosuppressed patients liable to contract infections, and worsen the severity of what would be minor infections in an individual with a fully functional immune system. The potential of producing cell, tissue, or organ grafts that do not require immunosuppression is a real holy grail of stem cell research and this is why the issue of personalized cell culture, discussed in Chapter 3, is so important.

The third big problem with cell therapy is delivery. Cells are normally injected as a suspension into the region where the repair is required. In the case of bone marrow transplantation, the injection can just be made into the bloodstream since the haematopoietic stem cells (HSCs), that are the critical component of the graft, can travel in the blood to niches in the host bone marrow and become established there. Solid organs pose more problems. In experimental cell therapy of the heart or spinal cord, it is usual to inject cells as a suspension into the region of damage. But it is clear from animal experiments that most cells die soon after injection and the objective of obtaining an integrated and functional three-dimensional arrangement of cells is far from being reached.

So cell therapy is currently stymied by several critical problems, especially the difficulty of obtaining enough of the required cell types, the problem of immunity and graft rejection, and the problems associated with effective delivery. These are all difficult issues, but they will be solved. The importance of stem cells is that they offer potential solutions to at least the first two problems. The cell supply issue can be addressed because pluripotent stem cells (ES cells and iPS cells) are capable of being grown in tissue culture without limit, and are capable of developing into any of the cell types normally found in the body. The problem of immunity and graft rejection could be addressed either by creating banks of cells such that there is always a reasonably good HLA match available for any individual, or by using one of the methods for producing personalized iPS cells. These are pluripotent stem cells produced from the patient that are a perfect immunological match and therefore provoke no graft rejection from the patient. The third problem, of effective delivery, lies less in the area of stem cell research and more in the area of bioengineering. The key is to improve the sophistication of tissue culture such that cells can be grown into particular three-dimensional shapes or tissue configurations, which can then be grafted as an integrated whole rather than being injected as individual cells.

Uncertain effectiveness of aspirational stem cell therapy

In addition to the examples of current stem cell therapy that will be discussed in this book, there is currently a great deal of other 'stem cell therapy' offered that does not have a clear scientific rationale. This usually involves autologous grafts of cells from one part of the body to another, or sometimes allogeneic grafts from some favourite cell line into the affected region. This type of procedure is rarely part of a controlled clinical trial. For the purposes of this book, this sort of activity will be called *aspirational stem cell therapy*, although it is also know as 'stem cell tourism' (Box 2).

Box 2: Stem cell tourism

For better or worse, we live in a world where money can be acquired simply by attracting attention and by making promises. Naturally, any promise of new cures for progressive and debilitating diseases is bound to attract attention. Partly because of genuine misunderstanding of the science; partly because of the very different outlook of clinicians and scientists, partly because of the ethical debates, and partly because of the desire to make money, the promise of 'stem cell therapy' has been made very loudly and frequently, and has attracted rather more attention than it really deserves. This has encouraged many thousands of people with severe, distressing, and debilitating diseases, often terminal ones, to seek a miracle cure from stem cell therapy. Such people are willing to travel across the world and pay large sums to clinics that are offering such cures. Indeed, why would people not do this?—it is literally a matter of life and death, and some hope of life from an experimental new treatment is always worth a gamble compared to inevitable deterioration and death. Unfortunately, most of what is on offer at present is almost certainly ineffective.

The tragedy of stem cell tourism arises from a bias familiar to all medical statisticians. In most diseases there is a lot of individual variation in the rate of progression, and it may be very difficult to predict what will happen to a particular individual in the next two or three years. Let us imagine a degenerative brain disease from which there is no recovery but where there are unpredictable periods of remission, and periods in which the patients feel better even if neuronal function is not actually improved. Suppose 100 such patients sign up for Dr Feelgood's miracle stem cell therapy. Purely by chance, some of these patients will experience a period of remission soon after they have had the treatment. They will feel extremely positive about this. Their gamble on the new therapy, and their considerable expenditure, appears to have paid off. These people are easily persuaded to write glowing testimonials praising Dr Feelgood and his cure, and these letters will soon find their way onto the clinic's website in order to attract more customers. Equally by chance, many others of the 100 patients did not improve, or got worse, shortly after the treatment. But they did not write testimonials, and if they wrote letters of complaint, they were politely reminded of the small print in the agreement which stated that the treatment was experimental and cure was not guaranteed. In any case, their letters did not get displayed on the clinic's website. This process, involving natural variability in the course of the disease and selective bias in the reporting of results, all too easily builds up an edifice of evidence for effectiveness which is entirely spurious.

Dr Feelgood probably does not have the contacts or resources to participate in a controlled clinical trial, so even if he wanted to establish efficacy by statistically acceptable methods he would not be able to do so. He is much more likely to believe the evidence of his own eyes: especially those grateful patients who receive his therapy and who improve shortly afterwards. This is known in the trade as 'anecdotal' evidence, and until the invention of the randomized controlled clinical trial in the 1940s, all judgement of efficacy in medicine was based on such unreliable methods.

It is rarely possible to establish effectiveness of a new treatment without a controlled clinical trial. Clinical trials are a very complex subject, but in essence they seek to avoid bias by making a comparison between one treatment and another which are both carried out in an identical manner on a whole group of patients. The ease of making the comparison depends inversely on the inherent variability of the disease. The more variable it is, the larger the test group needs to be in order to have the ability to detect a real difference in effectiveness of the treatment. Because the outcomes are often somewhat subjective it is important, if possible, that the patient does not know whether he is a member of the test group that receives the new treatment, or of the control group that does not receive it. It is even better if the medical staff who deal with the patients and record the outcomes also do not know who is in which group. There are many other complex problems surrounding clinical trials, but even meeting just the conditions of having a large enough sample to detect a difference, and of hiding the identity of the test and control groups, means that clinical trials can only be carried out by large academic medical centres, and often in several cooperating centres.

Most aspirational stem cell therapy is delivered by individual physicians or by private clinics which are not in a position to carry out controlled trials. Because patients are constantly asking for advice about such therapies, the issue has been very vexing for stem cell scientists. In 2008, the International Society for Stem Cell Research (ISSCR) came near to declaring that no new stem cell treatment should be carried out at all on human patients unless it was part of a controlled trial. Eventually the ISSCR did not adopt this very purist position and ended up declaring that new treatments needed to have some 'rationale and expectation of success'. This may seem obvious, but once again it raises the very different perception of the same data from the standpoint of different professionals, each embedded in their own culture. A rationale to a scientist is often quite different from a rationale to a clinician. For example, if injection of the patients' own bone marrow into the heart causes a two per cent improvement in function, but all the cells are thought to die within the first day, does this constitute a rationale for further treatment? The scientist will say 'no—we do not understand what is happening. We need to go back to the lab and find out before offering any more treatment.' On the other hand, the clinician will be inclined to say 'yes—two per cent is not much but it is better than nothing, and I must offer my patients the best. I really do not care what the mechanism is so long as there is some benefit.' This exchange represents just two of the cultures engaged with stem cells and viewing them from very different standpoints.

Chapter 2
Embryonic stem cells

Despite that fact that they are only just beginning to contribute to any form of therapy, embryonic stem (ES) cells are the cells about which all the fuss is made. More precisely, the fuss is made about *human* embryonic stem cells. ES cells were originally prepared from mouse embryos, and human ES cells followed some time later. Up until now mouse ES cells have actually been of greater practical importance than those of humans because they have enabled a substantial research industry to develop based on the creation of genetically modified mice.

All animal embryos start life as an *egg* fertilized by a sperm. The fusion product of egg and sperm is called a fertilized egg. This initially divides to form a small clump of similar cells. A typical embryo, such as that of a frog, will proceed to become a blastula (a ball of cells containing a cavity), then undergo a complex process of cell movement which results in the formation of three layers of cells called *germ layers*. The outer layer is the *ectoderm*, later to become the epidermis and the central nervous system. The middle layer is the *mesoderm*, later to become muscles and *connective tissues* (skeleton, tendons, sheaths around other structures); and the inner layer is the *endoderm*, later to become the epithelial lining of the gut and respiratory system. The cells in each layer then start differentiating to form a recognizable head, trunk, and tail, accompanied by the formation of cell clumps representing the precursors to each specific body part and organ. Although this process is extremely complex, it has become reasonably well understood at a molecular level through the developmental biology research of the last 30 years.

Because mammals bear live young, the course of early mammalian development differs in one main respect from that of the frog. After a few days the embryo becomes implanted in the uterus of the mother and is nourished by a placenta, through which nutrients pass from the maternal blood stream into that of the embryo or *foetus*. Part of the placenta develops from the tissues of the uterus, but part also develops from the embryo itself. Because the mammalian embryo produces both an actual embryo and also part of the placenta, it is more correctly referred to as a 'conceptus' than as an 'embryo'. However such is the emotive power of the word 'embryo', that attempts by scientists to use 'conceptus' or 'pre-embryo' have failed to catch on. The term 'embryo' is used here because of its greater familiarity, although 'conceptus' is certainly more appropriate.

A mammalian embryo has a few days of free-living existence before it implants in the uterus, during which it is known as a *preimplantation embryo* (Figure 3). This stage is a ball of cells, most of which are destined to become parts of the placenta. In fact, the early events of mammalian development are mostly concerned with making the placenta. In the first step, the outer layer of cells in the original ball of cells becomes a skin-like structure called the trophectoderm. Within the ball forms a fluid-filled cavity, and a clump of undifferentiated cells called the *inner cell mass*. At this stage the whole embryo is called a *blastocyst*, which is the mammalian version of the blastula of free-living embryos such as those of the frog. In normal development the inner cell mass produces several further layers of cells which are destined to become parts of the placenta, as well as those cells that form the actual embryo. Diversification of parts within the

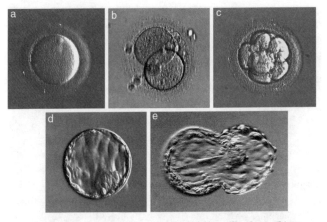

3. Preimplantation stage human embryos. (a) Fertilized egg. (b) Two cell stage. (c) Eight cell stage. (d) Blastocyst stage. The inner cell mass is the small clump of cells at bottom right. (e) Blastocyst in culture. Cells are moving out and adhering to the plastic surface

Courtesy of Kim Stelzig and Meri Firpo, University of Minnesota Stem Cell Institute

embryo commences with the formation of a *primitive streak* of cells at about 6 days of development in a mouse, or about 15 days in a human. This is the stage at which the three germ layers form, and it is also the stage beyond which twinning ceases to be possible. Before the primitive streak stage, identical twins can be generated by mechanically splitting the embryo of non-human mammals into two. On the basis of the arrangement of placental membranes in human twins, it is believed that the same is true for the spontaneous twinning in humans. The primitive streak is also the stage at which, for the first time, there is an identifiable population of cells that are solely destined to form the embryo proper (comprising a somewhat larger region than the primitive streak itself), rather than the placenta. For all these reasons, the primitive streak stage is regarded as the cut-off stage for experimentation on human embryos in the United Kingdom and in some other jurisdictions.

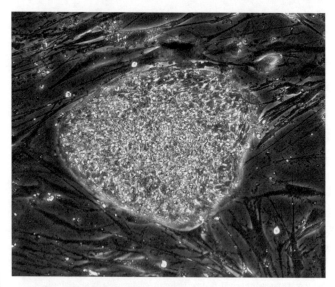

4. A colony of human ES cells in culture. The colony contains a few hundred small cells and is surrounded by a feeder layer of large elongated fibroblasts

Courtesy of Lucas Greder, University of Minnesota Stem Cell Institute

Embryonic stem (ES) cells are cells which are grown in tissue culture from the inner cell mass of a mammalian blastocyst-stage embryo. They are small cells that grow as tightly packed colonies. They are usually cultured on a layer of other tissue culture cells (*feeder cells*) which have been inactivated by irradiation by X-rays or drug treatment to prevent them dividing. This is why a picture of ES cells (Figure 4) usually shows a number of blobs (the colonies of ES cell themselves) surrounded by some ill-defined grey material (the monolayer of feeder cells). The feeder cells supply the ES cells with necessary growth factors and extracellular materials. However cultivation without feeder cells is also possible and, because of its convenience and the fact that cells destined for cell therapy must be grown in the absence of all animal-derived products, the methods for doing this will doubtless continue to be improved.

Mouse ES cells

Mouse ES cells were first isolated in 1981 by Martin Evans and Matthew Kaufman at the University of Cambridge, UK, and by Gail Martin at the University of California, San Francisco. They could be grown without limit and could differentiate into a wide range of cell types *in vitro*. In addition, they turned out to be very effective at colonizing mouse embryos into which they had been injected and contributing cells to all parts of the animal. These properties of mouse ES cells are shown in Figure 5.

There has been much debate about what cell type in the early embryo the ES cells actually correspond to. Is it the inner cell mass, which is actually used to establish the cultures, or is it some later stage of development? Examination of the gene expression patterns of ES cells suggests that they are somewhat different from all the normal cell populations within the embryo. Furthermore, there are some significant differences between mouse ES cells and human ES cells in terms of growth factor requirement, morphology, gene expression, and differentiation behaviour. These differences could indicate a different cell of origin or simply different changes evoked by the tissue culture environment.

Whichever cell type is the true *in vivo* counterpart of the ES cell, it is not a stem cell. All the early cell populations in the embryo are quite short lived and soon develop into other cell types committed to form specific body parts or tissue types. Because of this, ES cells have often been described as an '*in vitro* artefact'. However, this fact does not detract from their great importance for the various applications which we shall consider.

In an appropriate culture medium ES cells can be cultivated without limit. When plated without feeder cells in non-adhesive dishes, mouse ES cells will form *embryoid bodies* which are small structures in some ways resembling normal mouse embryos.

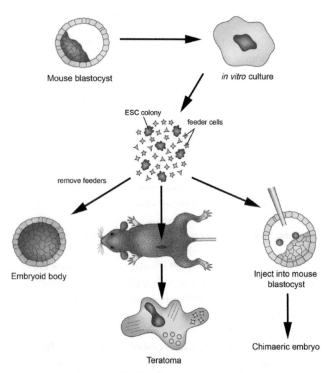

5. **Properties of mouse ES cells. They can grow in culture as undifferentiated cells and, under certain conditions, they can differentiate as embryoid bodies. They can be implanted into mouse early embryos, forming chimaeras. They can form teratoma-type tumours when implanted into adult immunodeficient mice**

However they are not the same as embryos. Firstly they are not enveloped by trophectoderm, the normal outer layer of the blastocyst. Secondly, they arise in a range of sizes and the different ratios of surface to volume and spatial relationships between parts means that they can be quite diverse in structure and composition. Over 2–3 weeks embryoid bodies normally generate cell populations representing many of the major body parts, tissue types, and cell types formed in a normal embryo. However, the spatial pattern is variable and abnormal and some tissue types, such as skeletal muscle, are rarely formed. Structures similar to embryoid bodies can also develop in a cell monolayer on adhesive plastic dishes.

When implanted into an adult animal with a compatible immune system, such as another mouse of the same inbred strain, mouse ES cells will form a type of tumour called a *teratoma* (Figure 5). This contains proliferating nests of cells similar to the original ES cells. It also contains zones of differentiation into many body parts and tissue types, whose arrangement is generally highly chaotic and varies from one tumour to the next.

The pluripotent character of ES cells is maintained by a network of *transcription factors*, which are proteins that bind to specific sequences in DNA and activate or repress expression of the adjacent genes. The most critical transcription factors for pluripotency are Oct 4, Sox2, and Nanog. These activate the genes for each other and repress genes required for the early steps of differentiation, resulting in a reasonably stable pluripotent state.

It was early shown that mouse ES cells could be injected into other mouse blastocysts where they would integrate into the inner cell mass of the host embryo and contribute to the development of all the resulting tissues. An organism formed from a mixture of genetically distinct cells, of which this is one type, is called a *chimaera*, recalling the chimaera of classical mythology which consisted of parts drawn from several different types of animal. Especially significant is the fact that ES cells can contribute to the germ cells of the chimaeric embryos: the sperm or the eggs. This means that it is possible to breed offspring from the chimaeric mice that carry the gene variants present in the ES cells that were injected at the blastocyst stage. The techniques for doing this were developed largely by Mario Capecchi at the University of Utah, and Oliver Smithies of University of Wisconsin, both of whom shared the 2007 Nobel Prize with

Martin Evans, one of the discoverers of mouse ES cells. The procedure enables the creation of virtually any desired line of genetically modified mice.

The importance of this cannot be overstated because it has been the basis of a vast amount of biomedical research in the last 25 years involving the use of tens of thousands of lines of genetically modified mice for all sorts of purposes. Briefly, these lines of mice have been made for the investigation of normal gene function and for the creation of 'mouse models' of human disease. Gene function is usually investigated by making a 'knockout' mouse, which completely lacks the gene of interest. The altered properties of such a mouse provides essential data for understanding what this gene does in normal development. Mouse models of human disease can be produced in which the same mutation is generated as is responsible for the human disease. This enables studies of the molecular pathology and the testing of new types of therapy, all on the small and relatively cheap scale of the mouse. A new drug that is shown to work on the mouse model will then be a candidate for testing in larger animals and eventually in humans.

It is precisely the two key properties of ES cells that have enabled this vast industry of genetically modified mice to exist. The ability to expand cells in tissue culture without limit is a precondition for effective genetic manipulation. When a new version of a gene is introduced into the cells, it is necessary to select those few cells in which it integrates into the host cell DNA in exactly the correct position. Such selection is only possible in tissue culture, where one modified cell out of many millions can be isolated and expanded using selective growth conditions. Mouse genetic manipulation also depends absolutely on the pluripotency of the ES cells, which means that they can enter the germ line following injection into host embryos, become transformed to eggs and sperm in the resulting chimaeric mice, and thereby enable breeding of new mice carrying the modified genes.

Human ES cells

Cells grown from inner cell masses of human embryos, sharing many properties with mouse ES cells, were first isolated in 1998, by James Thomson at the University of Wisconsin. This work was preceded by extensive work with non-human primates.

The reason that human ES cells have generated controversy is for one reason and one reason alone: they are made from human embryos. Just like mouse embryos, the first visible differentiation of the human embryo involves the formation of an outer trophectoderm surrounding a fluid filled cavity on one side of which is the inner cell mass (Figure 3). If the inner cell mass cells are taken out, provided with a suitable medium, and cultured on feeder cells, they will generate a line of human ES cells. There are certain differences in gene expression, appearance, and behaviour between mouse and human ES cells, but in general they are quite similar. They both share the two critical properties of being able to grow without limit and to generate a multiplicity of different differentiated cell types. We do not know whether human ES cells can integrate into a human embryo to form a chimaera. Such an experiment would be highly unethical because it might lead to the formation of a human baby with unpredictable abnormalities, and nobody would venture to attempt it. However, we do know that human ES cells can form embryoid bodies *in vitro* and that they can form teratomas when grafted to immunodeficient mice. Both embryoid bodies and teratomas generate many different tissue types and this indicates that human ES cells probably do share the high level of pluripotency shown by mouse ES cells.

The teratoma assay is particularly important for testing new lines of human ES cells to establish their pluripotency, since it is not ethical to implant into human embryos for this purpose. The cells are judged to be pluripotent if the teratomas to which they give rise form tissues characteristically derived from all three of the embryonic germ layers: ectoderm, mesoderm, and endoderm.

The mouse hosts for teratomas need to have a severely impaired (*immunodeficient*) immune system, otherwise the graft of human tissue would be rejected by the immune system of the host.

The word pluripotent has now become the standard term for describing the ability of ES cells to form a broad range of cell types. ES cells used to be described as 'totipotent', but this useage was discontinued because mouse ES cells do not normally form any trophectoderm when allowed to differentiate *in vitro*. Ironically, human ES cells, which are often thought to correspond to a more mature cell type *in vivo* than the inner cell mass, do form trophectoderm *in vitro*. Perhaps because there is no known rationale for this fact, the term 'totipotent' is also not used for human ES cells, and has become reserved just for the fertilized egg itself.

From where do the embryos come for the production of human ES cells? They come from *in vitro* fertilization (IVF) clinics. IVF is now a huge industry and there are clinics everywhere across the rich- and middle-income world. Normally the mother receives a hormone treatment which causes several of her *oocytes* to mature simultaneously, instead of the normal one per month. So as many as 10–12 oocytes may be harvested from one cycle, fertilized, and allowed to develop for a few days as preimplantation embryos. It is now considered unethical to implant more than two embryos at a time because of the risk of multiple pregnancy, so the remainder will be frozen for future use. If the first implantation is not successful, two further embryos can be thawed and reimplanted without the need for another cycle of hormone stimulation and oocyte harvest. But very often not all the frozen embryos are used. The parents may succeed with their pregnancy or pregnancies and not want any more, or they may decide for some other reason not to have further rounds of implantation. So all IVF clinics have large numbers of embryos stored in liquid nitrogen. At some stage the surplus embryos have to be disposed of. One option is simply to discard them. Another is to donate them for research purposes, including the establishment of new ES cell lines, and many parents are happy to do this.

Applications of human ES cells

In terms of potential practical applications, the overwhelming importance of human ES cells is that they offer a potential route to making the kinds of cells needed for cell therapy. We shall return later to this theme with examples related to specific diseases, and for the moment just consider the general principles of how it can be done. The secret is the understanding of normal embryonic development which has been achieved over the last 30 years through the efforts of developmental biology researchers. Embryonic development used to be completely mysterious: how can a simple ball of similar cells develop over a few days or weeks to become a miniature animal containing many groups of cells each committed to form a particular organ or body part? Through a combination of microsurgical, genetic, and molecular biology experiments we now know that embryonic development proceeds through a series of signalling and response events in each of which a small group of cells is exposed to a graded concentration of an extracellular substance. Cells above a threshold level are programmed to develop in one direction through the activation or repression of particular genes, those below the threshold concentration are programmed to develop in another. In the context of the embryo the extracellular signal substances are often called *inducing factors* or 'morphogens', but they belong to several of the same classes of secreted proteins as the growth factors and cytokines active after birth. A succession of such events brings about the formation of a complex body plan containing many organs and tissue types, starting from the simple blastula or ball of cells. Developmental biologists have now identified most of the inducing factors and also many of the genes that need to be turned on or off to achieve each step of the process.

The strategy for causing ES cells to become a particular cell type is therefore conceptually very simple. The cells are

Stages of induced development:

Endoderm →	Gut tube →	Posterior foregut →	Pancreatic bud →	Endocrine cells

Substances used in culture:

Activin Wnt	FGF10 Cyclopamine	FGF10 Cyclopamine Retinoic acid	DAPT Exendin 4	IGF1 Exendin 4 HGF

Media used for culture:

RPMI, RPMI+0.2%serum	RPMI 2% serum	DMEM 1%B27	DMEM 1%B27	CMRL 1%B27

Maxium duration of stage:

4 days	4 days	4 days	3 days	3 + days

6. An example of a procedure used to control differentiation of human ES cells, in this case towards insulin-producing pancreatic beta cells. There are five stages to the procedure, each involving a different culture medium and the addition of different growth factors and inhibitors to control the next developmental decision. This procedure was devised by the Californian company Viacyte Inc. (formerly Novocell Inc.)

exposed to the same sequence of inducing factors, at the same concentrations and times, as they would normally experience during embryonic development. Typically this involves four to six steps of treatments and each response from the ES cells makes them competent to respond to the next treatment. The final result is a population of the desired type of differentiated cell. In reality, the normal events are not understood with complete precision, the environment of the culture dish is somewhat different from that of the intact embryo, and different lines of ES cell behave slightly differently. For all these reasons there is some need for empirical testing as well as rational design, and different laboratories may produce slightly different protocols. However the essential strategy is common to all labs and may be illustrated by the example shown in Figure 6, which shows a protocol for the production of insulin-producing cells from human ES cells.

Although the public considers that cell therapy is the obvious application for human ES cells, it is worth noting that many of the scientists who work with human ES cells are often somewhat sceptical that they will actually be used for therapy. They argue that the importance of human ES cells lies in other areas, three of which are currently envisaged. First, there is the investigation of normal development. The availability of cells that will carry out some developmental steps *in vitro* offer a method for investigating certain aspects of normal human embryonic development which would not otherwise be accessible because it would not be ethical to experiment directly on human embryos after they have implanted in the womb. Secondly, there is the study of cellular pathology for those human genetic diseases where the relevant cells can be obtained *in vitro*. While tissue samples may sometimes be obtained from affected individuals, ES cells carrying a genetic disease mutation give access to those embryonic and immature cell populations whose function may be compromised as the primary effect of the mutation. Their availability increases the potential for studying the pathological processes involved in the disease. Thirdly, there is the possibility of obtaining normal or genetically abnormal human cells for drug screening. Some cell types, such as heart muscle (cardiomyocytes), are very difficult to obtain in viable form, and even human liver cells (hepatocytes) are in short supply. The ability to make any quantity of such cells will transform the ability to screen potential drugs and will probably reduce the number of animals needed for this purpose.

Probably all of these things will happen and they may prove quite important. But they are niche activities and the truth is that the public really wants cures and especially in those areas where a lot of public money has been invested in human ES cell research, it wants them fast. At the same time, a minority of the public, particularly in the USA, are strongly opposed to any use of human embryos or ES cells for any purpose. The total incompatibility of views between those who want cures and those who want protection for frozen embryos is the cause of the long and intense ethical debate on this matter (Box 3).

Box 3: The ethical debate about human ES cells

The opponents of stem cell research consider that human preimplantation embryos, such as are depicted in Figure 3, should have full human rights and that to use them to make ES cells is tantamount to murder. In the USA the colourful expression 'miniature Americans' has been used to describe frozen preimplantation embryos. There is a programme run by the adoption organization Nightlight to promote the 'adoption' of the 400,000 embryos that are 'frozen, waiting to be chosen', in other words make them available for embryo transfer to willing surrogate mothers. Most people who think like this are very religious and they often consider that they are following the teaching of their faith in their opposition to stem cell research. The Catholic Church is indeed hostile to the use of human ES cells for any purpose, as are some Protestant groups. They consider that human life begins at fertilization and that fertilized eggs and all later developmental stages should have full human rights.

Ironically, in the Middle Ages, before the discovery of the human sperm and eggs, the Catholic Church taught that the soul entered the human foetus at the time of quickening, when the mother can first feel the foetus move inside her. But this is about 18–24 weeks of gestation, well after the time at which most abortions are performed, and even longer since the foetus was a preimplantation embryo, so this teaching has been quietly forgotten.

As far as other religions are concerned, the Buddhist view is similar to the Catholic, considering personhood to begin at conception. According to Jewish and Islamic teaching the human embryo has no special status until 40 days have passed, following which personhood develops gradually. Jews in particular tend to be supportive of research using human ES cells. Hindus consider that personhood depends on the reincarnation of the former person, which is believed to occur, depending on the group within the faith, sometime between conception and 7 months of gestation.

So the truth is that there is a considerable diversity of views about this issue among the adherents of different religions.

What do scientists think? Individuals may have different opinions on all sorts of things, but the vast majority of biomedical scientists consider that personhood develops gradually and that preimplantation embryos are not the same as human beings, being more akin to cultured human cells or tissue samples. For example, the blood bank is a familiar example of something that is genetically human and is alive, but is not a human being. The same could be said of donor organs, or tissues, or other types of human cell grown in culture. Such cells or tissue samples are subject to various regulations concerning informed consent for their use. There have been various debates about how to frame these regulations, and the occasional scandal arising from misunderstandings, but nobody has ever claimed that human cells in tissue culture, or donor kidneys, or frozen white blood cells, or other types of entity that are both human and alive, should be accorded the same sort of human rights as intact and self-conscious human beings. Given that human ES cells are grown from surplus preimplantation embryos donated by the parents, and that they would otherwise have to be discarded, very few scientists are opposed to the practice.

Chapter 3

Personalized pluripotent stem cells

Stem cells and cloning

The creation and use of lines of human ES cells has generated considerable debate because of their origin from human embryos. But stem cell biology is also involved with another issue that raises a red flag to a different but overlapping set of opponents. This is *cloning*, and the connection to stem cell biology lies in the potential routes that have been sought to make personalized pluripotent stem cells that are a perfect genetic match to a specific patient, and will therefore provoke no immune rejection on grafting.

To clone something means to make a set of identical genetic copies. You can clone molecules of DNA by taking one molecule and putting it in a microorganism so it is copied to make millions of identical molecules. Molecular cloning is carried out in every biomedical laboratory around the world virtually every day. It caused a brief flare of controversy when the techniques were first introduced in the 1970s, mostly because of exaggerated fears about potential safety issues, but it is now a routine activity on which the whole of biotechnology depends. You can also clone cells in tissue culture. If you take one cell and grow it to form a colony then this is a clone, as all the cells in the colony are genetically identical to the founder. Cellular cloning is also practiced for various purposes in numerous labs around the world every day and is not a subject of controversy. But to the lay person 'cloning' does not refer to molecular or cellular cloning, it means making whole animals which are exact genetic copies of another animal. If you can do it with animals then it is a short technological step to doing it with people, and the spectre of human cloning is easily conjured up.

Animal cloning is deeply rooted in developmental biology, which is the science that studies how embryos work at the cellular and molecular level. The first clones were made in the late nineteenth century by separating cells of frog and sea urchin embryos and rearing these early embryo cells in isolation. In many types of animal, including mammals, the two cells resulting from division of the fertilized egg will each form a complete embryo. The embryos are half normal size, but it will catch up with their full-size siblings at a later stage when they start to grow. Such clones are effectively identical twins, and it is believed that most human identical twins arise by this very mechanism, separation of the first two cells formed by division of the fertilized egg.

The most famous name in embryology during the first half of the twentieth century was that of Hans Spemann, who worked at the University of Freiburg in Germany. He received the Nobel Prize for Physiology in 1935 for his discovery of embryonic induction, the signalling and response process that underlies animal (and human) development. One of his experiments represented the first example of the cloning of an animal embryo using a nucleus from a later stage than the first two cells. He showed that if a fertilized newt egg is constricted to a dumb-bell shape by a fine ligature, then the half containing the nucleus divides while the other half does not. Following a few cell divisions, relaxation of the ligature may allow a single nucleus to pass into the uncleaved half from a cell on the cleaved side. Despite the fact that this nucleus has already divided several times, Spemann showed that it could support development of the uncleaved half egg and cause it to give rise to a complete embryo.

Developmental biologists then wondered for how long this would remain true. Robert Briggs and Thomas King, at the Institute for Cancer Research in Philadelphia, transplanted nuclei from the blastula stage (ball of simple cells) of frog embryos into enucleated eggs. They found that complete embryos could be obtained. In subsequent experiments, they examined the ability of nuclei that were taken from later developmental stages to support development of enucleated eggs. The results showed after the embryo had progressed beyond a simple ball of cells, the results became progressively less good. These experiments were developed further by John Gurdon, at Oxford and later Cambridge University. He obtained similar results to Briggs and King but laid stress on the very small proportion of transplants where late stage nuclei did produce reasonable quality tadpoles. He also proved that nuclei from differentiated skin keratinocytes could, with very low efficiency, support the development of tadpoles, although not of mature frogs.

By the 1950s the structure of DNA had been established. It was known that genes were made of DNA and it was believed that every cell in the animal contained the same set of genes. So the ability of a differentiated cell nucleus to support development of an egg was the expected result, and it is probably because it was the expected result that Gurdon's very low success rate was greeted as a positive rather than a negative result. Ever since these experiments with frog embryos it has been generally accepted that all the genes are present in at least most cell types throughout development, so successful cloning by nuclear transplantation is just a matter of 'reprogramming' the nucleus to reset the program of gene expression back to that characteristic of the fertilized egg.

Animal cloning really hit the headlines in 1997 when a similar technique to that developed by Briggs and King was used to make Dolly the sheep. The originator, Ian Wilmut, worked at an agricultural research station, the Roslin Institute, near Edinburgh, Scotland, and had been principally interested in improving the methods for introducing genes into farm animals. He was able to take a nucleus from a sheep tissue culture cell and transplant into the enucleated oocyte of a ewe. Since sheep are mammals, the resulting embryo needs to be transferred back into the uterus of another ewe who serves as a surrogate mother. Only a very small proportion of the nuclear transfer embryos developed, but one was born and survived to achieve media immortality as Dolly. Over the next few years, many other species of mammal were cloned by similar methods, including mice, rats, cats, hamsters, pigs, and cattle (Figure 7). The method is now known as *somatic cell nuclear transfer (SCNT)*, a *somatic* cell being any cell that is not a germ cell.

Dolly caused a huge media furore because sheep, being mammals, are so much closer to humans in their reproductive biology than are frogs. It made the possibility of human *reproductive cloning* seem very near. In fact no human reproductive cloning has ever happened and virtually all scientists are opposed to it for reasons of safety. Studies of gene expression in animal embryos that have been made by SCNT have indicated many abnormalities, and it is highly likely that some of these would result in anatomical abnormalities leading to handicap in a newborn child.

Therapeutic cloning

Everyone can agree, albeit for different reasons, that human reproductive cloning at the present time would be a bad thing. But there is another very important capability possessed by an early embryo made by nuclear transfer: it can be used to establish an ES cell line. Such a cell line will be genetically identical to the donor of the original nucleus, and because ES cells can be used as a source for several therapeutically useful cell types, the cell line could serve as a source of immunologically compatible grafts for that person.

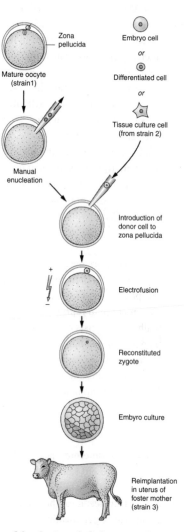

7. Procedures used for cloning of whole mammals. The resulting animal will have the genetic constitution of the cell nucleus that was introduced into the oocyte

38

Because this procedure involves no formation of an actual cloned individual, it has become known as *therapeutic cloning*.

However, human therapeutic cloning has never been successfully achieved (see Box 4). Experiments with many species of animal have all shown that somatic cell nuclear transfer is a very inefficient process. Similar to the original frog embryo nuclear transfers done at late stages, only a small minority of reconstituted eggs are capable of developing even far enough to establish an ES cell line. This makes it a problematic procedure to use in humans because of the difficulty of obtaining human oocytes, which are the essential starting material. These can only be obtained from female volunteers who undergo the hormonal stimulation and laparoscopic harvest of oocytes in the same way as is done for IVF. This procedure is not pleasant and involves some risk. Some women are willing to do it for money, and they are the egg donors to IVF clinics for the type of fertility treatment involving egg donation. But research institutions are not usually permitted to pay donors and so they receive no oocytes.

Induced pluripotent stem (iPS) cells

It is now unlikely that human therapeutic cloning is ever going to be practised, partly because of the ethical issues, partly because of the difficulty of obtaining oocytes, and partly because a much easier method for making pluripotent cells has now been invented. *Induced pluripotent stem cells* (*iPS cells*), which are extremely similar to ES cells, are made by introducing a few specific genes into normal cells. This technique was a remarkable discovery made by Dr Shinya Yamanaka of Kyoto University, Japan, in 2006. What he did was to introduce a large set of genes, known to be important for ES cell function, into fibroblasts, and look to see if any ES-like colonies appeared. Then each gene in turn was subtracted from the mixture to find the ones that were really essential. Then the essential ones were tested again in all possible combinations. The result was a set of four genes—*Oct*4,

Box 4: Human cloning

Human *reproductive cloning* is theoretically possible by the same means as used for animal cloning: somatic cell nuclear transfer (SCNT) followed by reimplantation of the cloned embryo into the uterus of a surrogate mother. This is strongly opposed by virtually all scientists on safety grounds because animal experiments indicate that there will be a high risk of abnormalities. Ethicists, philosophers, and religious authorities seem to be less interested in safety but tend to feel that cloning would violate 'human dignity' although there seems to be little agreement among them about what 'human dignity' actually means, as was made clear when a definition was sought by President Bush's Council on Bioethics.

The main demand for human reproductive cloning, were it ever to become practical, is not from dictators who wish to create vast armies of loyal automatons (these are readily available anyway), but from infertile couples who are desperate for a child and will do anything to get one. It is an interesting issue for the future whether human cloning would eventually become acceptable for fertility treatment were it actually to be an efficient and safe procedure.

Human *therapeutic cloning*, the use of SCNT embryos to create ES cell lines, has been strongly opposed in some countries, such as Ireland, Poland, Italy, and Germany, mostly by religious groups. The reasons are that it involves 'destruction' of human embryos in the same way as the establishment of ES cells from normal embryos, and also that it is supposed to violate 'human dignity'. Because therapeutic cloning is regarded as 'worse' than simple ES cell derivation, there is also a group of countries that permit the latter but not the former, including Brazil, Canada, France, and Iran. In a third group of countries, including the UK, Sweden, China, India, and Australia, therapeutic cloning is permitted for medical research, under various forms of licence and regulation.

Despite the heated debate on this subject, no human ES cell lines have ever actually been created using SCNT. In 2004–5, much publicity was accorded to two papers in the journal *Science* from the lab of Dr Hwang Woo Suk in Seoul, South Korea. The first of these claimed to have produced an ES line from a reconstituted human embryo with a somatic nucleus, the second to produce a variety of ES lines from several donors carrying different genetic diseases. The initially uncritical acceptance of this work was certainly because it was the expected next step. However, the results were later shown to be incorrect and the papers were retracted. The original cell line was found to be parthenogenetic, meaning that it was actually derived from an unfertilized egg whose nucleus had not been properly removed.

Sox2, *Klf4*, and *c-Myc*—which could generate about one ES cell colony per 10,000 input cells. The protein products of *Oct4* and *Sox2* are in the core group of pluripotency-regulating transcription factors that are normally active in ES cells. *Klf4* and *c-Myc* were already known as *oncogenes*, genes that can cause cancer if activated inappropriately; in iPS cell generation their functions are not very well defined but they do considerably increase the efficiency of the overall process.

This work set off an explosion of activity around the world and the technology of preparing iPS cells has since advanced very rapidly (Figures 8 and 9), with human iPS cells being reported in 2007, only one year after the initial discovery (Figure 10). It is now known that various other genes can be substituted for *Klf4* and *c-Myc*, and various ways have been found to increase the efficiency of the process still further.

Mouse iPS cells were soon shown to be capable of contributing to all tissues in mouse embryos including contribution to the germ line, and thus really were like ES cells in this crucial respect.

Stem Cells

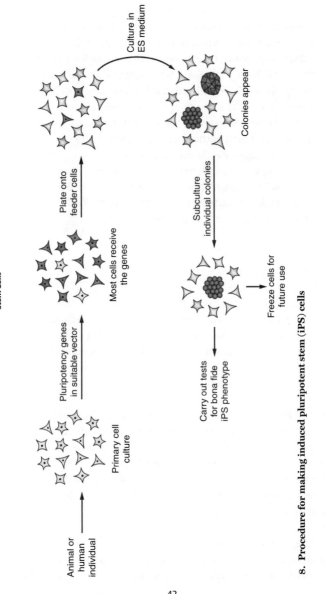

8. **Procedure for making induced pluripotent stem (iPS) cells**

Animal or human individual

Primary cell culture

Pluripotency genes in suitable vector

Most cells receive the genes

Plate onto feeder cells

Culture in ES medium

Colonies appear

Subculture individual colonies

Carry out tests for bona fide iPS phenotype

Freeze cells for future use

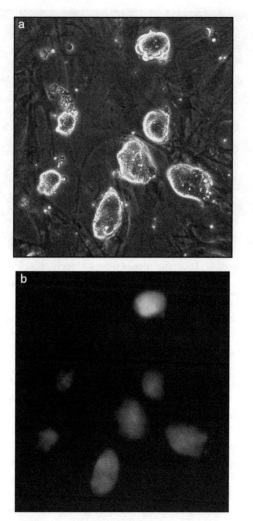

9. Mouse iPS cells. (a) Colonies of iPS cells growing on a fibroblast feeder layer. (b) The same field of view visualized to show expression of the endogenous *Oct4* gene of the cells

Courtesy of James Dutton, University of Minnesota Stem Cell Institute

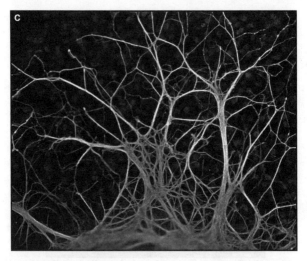

9. (continued) (c) Differentiation of neurons (nerve cells) from this cell line

It was also found possible to make iPS cells from cell types other than fibroblasts. In particular, white blood cells can serve as starting material, a cell type which is very easy to obtain from any human individual.

The genes are delivered using special viruses that insert their DNA into the chromosomes of the cells and enable high level expression of the products. The proportion of treated cells that become iPS cells is very low because a large number of molecular-scale events need to occur in the correct way to achieve complete reprogramming to the pluripotent state. This does not happen in most of the treated cells so a selective culture system is necessary to isolate the few fully reprogrammed iPS cells from the many unreprogrammed or partially reprogrammed cells. Following infection with the gene-carrying viruses, the cells are plated onto feeder cells in ES cell culture medium. Under these conditions ES cells grow fast whereas the parent cell type does not grow. Once colonies have appeared, those resembling ES

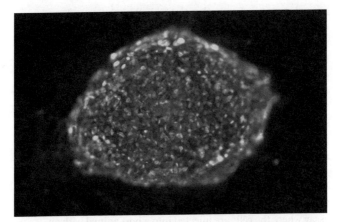

10. A colony of human iPS cells. These are visualized by fluorescent labelling of a cell surface molecule called TRA1-81, characteristic of human ES and iPS cells, so the feeder cells are not visible
Courtesy of Lucas Greder, University of Minnesota Stem Cell Institute

cells are picked and subcultured on fresh feeder cells. Eventually the viral-encoded genes usually become 'silenced', that is they cease to be active. Continued activity of the introduced genes is detrimental, because, being pluripotency factors, they will inhibit differentiation of the cells.

There is a set of standard tests which are used to characterize iPS cells, and to establish whether a particular new line has a genuinely ES-like phenotype or not. One key property of mouse ES cells is the ability to form chimaeras when injected into early mouse embryos. Ideally these should be germ-line chimeras, i.e. some of the donor cells become germ cells (sperm or eggs according to the sex of the animal) and the resulting mice should be capable of reproducing and generating offspring of the ES cell genotype. Even more demanding is the tetraploid rescue test. Here the host embryo is caused to become tetraploid (i.e. the normal chromosome number is doubled) by electric pulse-induced fusion of the first two cells into a single cell. For reasons that are not well understood, tetraploid cells cannot contribute to the foetus although they can still form placental structures. When good quality ES cells are injected into a tetraploid host embryo, they can form the entire foetus with no significant contribution from host cells. This has also been achieved with iPS cells, showing that it is possible to generate a complete animal from a pure population of such cells, given the environment of the early mouse embryo.

For human iPS cells it is not possible for ethical reasons to inject the cells into embryos, so the standard approach is the *teratoma* assay where the cells are injected into a host animal with an impaired immune system. Good quality iPS cells should grow to form a teratoma, and this should contain tissues characteristically derived from all three embryonic germ layers: ectoderm, mesoderm, and endoderm.

Patient-specific iPS cell lines

The properties of iPS cells can be very similar, or even identical to, ES cells. But the technology has an important practical attribute that ES cells do not share, which is the ability to establish new cell lines from specific individuals. For years there has been speculation about the possibility of creating patient-specific ES cell lines. But this has relied on the hope that it would be possible to do somatic cell nuclear transfer of patient cell nuclei into donor oocytes in order to create preimplantation embryos of a genotype identical to the patient. This would then be followed by derivation of an ES line from the embryos thus created. As discussed above, this procedure has never actually been successfully achieved using human cells. The main practical problems are the very low efficiency of cloning by SCNT coupled with the difficulty of obtaining enough human oocytes to make the experiments feasible.

In contrast to the difficulties surrounding therapeutic cloning, many patient-specific iPS lines have already been made, including many from people with specific genetic diseases. The technology appears robust and has been used in numerous laboratories. The attraction of the patient-specific cell line is that any differentiated cells made from it will be a perfect immunological match to the donor and can therefore potentially be grafted back without the use of immunosuppressive drugs.

Initially iPS cells were made from fibroblasts taken from a small skin biopsy. But many cell types can be used and methods have now been devised to make iPS cells from blood. Most of the cells in the blood are red cells that have no nuclei and are biologically 'dead'. But a blood sample also contains the white cells (lymphocytes), which do have nuclei. Although most of these white cells will not grow in culture, it is possible to stimulate lymphocytes so that they will grow for a few passages, and this period of expansion in tissue culture is sufficient to initiate the reprogramming to iPS cells. Because taking blood samples is quick, simple and almost painless it will probably become the starting material of choice for making iPS cells from any human individual. Because iPS cells are not made from embryos they are usually regarded as being free from ethical problems. However, there are always ethical issues and some are mentioned in Box 5.

In order to make clinical delivery of patient-specific cells a reality it will be necessary to find a routine way to make iPS cells without the need to insert genes into the cells' own DNA. This is currently done using viruses. But it is known that gene insertion can create mutations during the integration process because it may interrupt host genes, or cause host genes to be inappropriately activated. Moreover, the silenced genes encoding the pluripotency factors may become reactivated at low frequency and may subsequently cause the formation of tumours. This has been noticed in mice grown from iPS cells, especially when the *c-Myc* gene was one of those used to make the iPS cells.

Box 5: Ethics of iPS cells

The new technology of iPS cells has been greeted enthusiastically by opponents of human ES cell research. At first sight this may appear odd. If two flasks of cells have identical properties, how can one be 'good' and the other 'bad'? The answer again lies in the question of origin. If you believe that a preimplantation human embryo is a person with human rights, then cells made from such an embryo are necessarily bad, whereas iPS cells can be prepared from a small skin biopsy from a willing donor, or even the white blood cells from a small blood sample, and are therefore good.

iPS cells are described as 'adult stem cells' by the opponents of stem cell research even though they are essentially identical to ES cells and quite different from tissue-specific stem cells found in adults.

However, iPS cells are not devoid of ethical issues. We may soon see banks of iPS cells made from numerous living donors. Should the donors retain any rights in relation to the uses to which the cells are put, or to any commercial income that might accrue from them? Will the donors be protected against any discovery of adverse mutations in the DNA that might affect life prospects or life insurance? Moreover, human iPS cells could theoretically be injected into blastocysts to generate chimaeras, which could be another route to reproductive cloning.

These and many other questions will no doubt keep the ever increasing cohorts of bioethicists employed for many years to come.

However, there are other ways of introducing the pluripotency factors and encouraging results have been obtained with several methods. This is an area of rapid technical progress and in the future it is safe to assume that an efficient method not involving gene integration will allow routine generation of iPS cells from any individual.

An aspect of this that scientists generally do not think about is money. Most financial analysts are very sceptical about personalized cell culture, considering that the costs are too high for this to be the basis of an economically viable form of treatment. This is certainly true at the present time. But the history of technological innovation suggests that techniques which are initially very expensive can be made into cheap mass production procedures over a course of 10–20 years. Personalized cell culture is particularly attractive for a disease like diabetes, where patients can be kept in pretty good health for months while the cells are prepared, and where long term immunosuppression of grafts is undesirable in terms of the overall cost-benefit balance for the treatment.

Chapter 4

Potential therapies using pluripotent stem cells

Pluripotent stem cells may be either ES cells or iPS cells. Their properties are essentially the same, but iPS cells have the potential to be prepared from an individual patient and thus be a perfect genetic match to that patient. This chapter will examine some of the diseases that may be the first to be treated by cell therapy using pluripotent stem cells. At the time of writing, two clinical trials had commenced using human ES cells, one in the USA and the other in both USA and Europe.

The proposed cell therapies will in all cases involve making the required differentiated cells *in vitro* and then implanting them into the appropriate site in the patient. They do not involve implanting the pluripotent stem cells themselves because one of their properties is to form tumours called teratomas when implanted into a host. Like most tumours, teratomas can be very dangerous and the primary safety requirement for this type of cell therapy is to ensure that there are no pluripotent cells left in the cell population to be implanted, so that the danger of forming a teratoma is negligible. There are some other safety issues which also loom large in this area. Regulatory authorities such as the Food and Drug Administration (FDA) of the USA insist that cells for cell therapy should be cultured under conditions conforming to 'Good Manufacturing Practice (GMP)'. This is a very complex area, but essentially means that no animal products should be used in the manufacture, that all substances that are used are of a required degree of purity, that the facility satisfies criteria of restricted access, sterility and air handling, and that extremely detailed records are kept of every stage of the procedure. All this is very expensive and represents a big culture shift for academic scientists who are used to working in a much less regulated manner.

In all areas of clinical research the requirements of the FDA, and similar bodies in other countries, have become more and more onerous over recent decades and this is one of the reasons for the relatively slow progress of stem cell therapy. While irksome to researchers, the regulations do represent an inevitable trade-off between enabling some new treatments to be developed, and avoiding disasters from premature introduction of untried treatments. Without doubt they have slowed down the introduction of new treatments in recent decades, but they have also avoided a repeat of the thalidomide tragedy of the 1950s and 1960s, in which thousands of children were born with missing or defective limbs as a result of their mothers taking a new drug to control morning sickness in pregnancy. The FDA will accept a certain level of risk for a new treatment that is to be applied to a group of very sick patients who would otherwise die in the near future, but not for a group who have a relatively long life expectancy on the basis of existing therapies.

Diabetes

Diabetes is a common disease characterized by high and uncontrolled levels of blood glucose. The known number of patients worldwide is over 200 million and this is expected to double over the next 25 years. The most important event in the therapeutic history of diabetes was the discovery of *insulin* in 1921. This is the principal hormone that controls the blood glucose level and it is secreted from the pancreas. It was discovered at the University of Toronto by a surgeon, Dr Frederick Banting, assisted by a medical student, Charles Best. The availability of insulin introduced the present era when diabetes patients can expect long-term management of their disease rather than succumbing to an inevitable and rapid death. Largely because of insulin, as well as more recently introduced drug treatments, diabetes is sometimes believed no longer to be a serious medical problem. But health professionals and patients know otherwise. Diabetes remains a huge problem not least because it consumes something like 10–15 per cent of the entire health care budget of the wealthier countries in the world. Although most cases of diabetes can now be managed so that the patients expect a lifespan approaching normal, the inability to control blood glucose concentration with precision leads to gradually accumulating damage to blood vessels. In the long term, this damage causes complications which can be severe and distressing. They include heart disease, stroke, blindness, and peripheral vascular disease leading to difficulty healing minor wounds and ulcers and sometimes even to amputations. In addition, the self-discipline required to monitor blood glucose several times each day and to inject insulin accordingly is considerable, and the side effects of other drugs that are used to treat diabetes can be unpleasant. For all these reasons, there is a continued demand for a 'cure' for diabetes, not just a method of long-term management. In addition, diabetes is one of the top targets for cell therapy based on pluripotent stem cells because it can build on an existing form of cell therapy: islet transplantation, which is described below.

The key cell type when considering diabetes is the *beta cell* which resides in the *islets of Langerhans* in the pancreas. The pancreas is an organ in the abdomen which is mostly concerned with secreting digestive juices into the intestine. But about 2 per cent of its cells have a quite different function, being *endocrine cells* that secrete hormones into the bloodstream. Most of the endocrine cells are grouped into small clusters which were discovered by the German medical student Paul Langerhans in 1869, and named 'islets' by him. The islets contain several types of endocrine cell but the beta cells are the most numerous and, at least in terms of the consequences of their absence, the most important. Beta cells are the only cells in the body that produce insulin, and insulin is central to the nutrition and metabolism of several tissues, especially muscle and adipose (fat) tissue, because it is necessary to enable the uptake of glucose into cells. Under normal circumstances, a rise in blood glucose following a meal will stimulate the beta cells of the pancreas to secrete insulin and this will lead to uptake of the glucose by adipose tissue and muscle, and also by the liver where it is converted into glycogen and fats. In the absence of insulin, the concentration of glucose in the blood will rise uncontrollably and the patient will be unable to make use of it.

There are two main types of diabetes. Type 1 diabetes usually strikes during childhood or young adulthood and is characterized by a loss of the beta cells due to an autoimmune attack. The initial causes of the autoimmunity remain unclear but it is sometimes thought to be triggered by a viral infection as well as genetic predisposition. Although beta cells may regenerate to a limited extent, patients with type 1 diabetes will eventually lose all their beta cells and will then die rapidly without treatment. They are absolutely dependent on injected insulin and need to monitor their diet carefully and inject the correct dose of insulin regularly to achieve reasonable control of blood glucose.

Type 2 diabetes is more common and tends to develop at older ages. It is a complex and multifactorial disease, but usually seems to involve some pathology in the beta cells such that they cannot adapt to an increased demand for insulin, for example following development of obesity. There is also often an insulin resistance of the peripheral tissues such that the available insulin does not have a sufficient effect and cannot drive uptake of the excess glucose. Therapy for type 2 diabetes usually involves treatment with various drugs to slow the release of glucose from the gut, to enhance the secretion of insulin from beta cells, and to enhance the activity of the available insulin. Often, as the disease progresses, type 2 patients also take insulin by injection, usually once or twice per day, as well as continuing with their drug treatments.

Cell therapy for type 1 diabetes

In the last ten years, a partly effective form of cell therapy has been introduced in which islets taken from the pancreases of deceased human organ donors are grafted into diabetic patients. Its success depends on a refined procedure for immunosuppression called the Edmonton Protocol, after the University of Alberta, Edmonton, Canada, where it was devised. This therapy is mostly used for patients with unawareness of hypoglycaemia (low blood glucose). Although diabetes is associated with high blood glucose, a low level can also be dangerous, ultimately because not enough is available to maintain brain function and the patient can become unconscious. In severe diabetes the usual mechanisms for mobilizing glucose, based on another hormone called glucagon, may be impaired. Usually when blood glucose is too low, adrenalin is released and causes shakiness, sweating, anxiety, and hunger, but this response too can eventually be lost and then the patient is vulnerable to passing out suddenly while driving his car, and thus suffering a crash; or falling to the ground and suffering a head injury; or dying during sleep. For these reasons, unawareness of hypoglycaemia is really life threatening.

The islet transplantation technique involves taking islets from the pancreas of a deceased organ donor, and infusing them into the liver through the hepatic portal vein, which normally connects the intestine and liver. The islets will lodge in the liver and, because the blood supply to which they are exposed carries the nutrients that have recently been absorbed from the intestine, they should secrete appropriate amounts of insulin to manage the glucose levels that are encountered.

The therapy has proved quite successful. All treated patients do recover from their unawareness of hypoglycaemia and a proportion are able to stop taking insulin altogether. Unlike the aspirational types of stem cell therapy that will be mentioned later, islet transplantation really does work. This can be proved not just by asking the patients how they feel, but by actual measurements of blood glucose and of C-peptide, a by-product of insulin production which is released into the blood and whose concentration serves as an objective indicator of beta cell function.

But islet transplantation suffers from two major problems. First the supply of donor cells is hopelessly inadequate to meet the demand. This situation is common to all types of graft that are dependent on organ donors. Secondly these grafts, like any graft between different individuals, are *allografts* and are therefore subject to rejection by the immune system of the host. This means that the recipients have to receive immunosuppressive drugs for the rest of their lives. These drugs can cause unpleasant side effects and they also damage the delicate beta cells of the graft and reduce their lifespan and effectiveness.

The potential of stem cell research in this field is obvious. If beta cells could be made *in vitro* from pluripotent stem cells then the supply will become potentially infinite. Furthermore, if the stem cell source is a line of iPS cells made from the individual patient, then the cells will be a perfect genetic match and no immunosuppression will be required to suppress graft rejection. Actually the rejection issue is not quite so simple in the case of type 1 diabetes. Because type 1 diabetes is an autoimmune disease, the patient is likely to mount an autoimmune attack on the graft even if it is a perfect genetic match. This is not a problem for current clinical islet transplantation because the level of immunosuppression given is sufficient to suppress the autoimmunity as well as the alloimmunity, but it may be a problem in the future. There are some potential ways of dealing with autoimmunity which have been used in animal experiments, including the encapsulation of the cells in a material that will allow passage of insulin but not of cytotoxic lymphocytes from the host. Alternatively, a low dose, and therefore more acceptable, immunosuppression regime has been shown to work in whole pancreas grafts between identical twins, where the recipient is diabetic. In this situation there is no alloimmunity, but there is an autoimmune attack mounted by the host against the islets of the graft.

Making beta cells from pluripotent stem cells

The methods for making beta cells depend on a good understanding of normal beta cell and pancreatic development which has been achieved in the last 15 years through the efforts of developmental biology researchers. Starting from the primitive streak stage of the embryo, first the endoderm germ layer is formed, this then becomes subdivided and a foregut territory is established. The foregut produces two pancreatic buds which grow out from the primitive gut tube and subsequently fuse together to become a single organ. Within the pancreatic buds, endocrine progenitor cells develop. Some of these will become beta cells, and the rest will become the other endocrine cell types found in the islets of Langerhans. The protocols designed to direct differentiation of beta cells from pluripotent stem cells therefore involve five steps: to endoderm, to foregut, to pancreatic bud, to endocrine precursor, and finally to beta cell. Each of these steps is achieved by treatment with a specific inducing factor and success is monitored by detecting the activation of the key genes that are known to be required at each stage. Different labs have devised slightly different protocols but they are all similar in principle and an example was shown in Figure 6 above. The differentiation protocols are not yet perfect, as the proportion of beta cells produced falls well short of 100 per cent. Moreover, the cells are not fully mature. They rather resemble an immature type of beta cell found in the foetal pancreas which is not responsive to glucose. Mature, glucose-responsive beta cells do develop if even more immature cells corresponding to the stage of the pancreatic bud are implanted into animals, and current thinking is that these cells could also be implanted into human patients, encased in a semi-permeable material to prevent immunorejection.

Evidence of effectiveness of all proposed new treatments is obtained through animal experiments. In the case of diabetes it is possible to destroy the beta cells of a mouse by injection of a drug called streptozotocin. If immunodeficient mice are used they will tolerate grafts of human tissue, which are normally inserted into the kidney. It has been shown that relief of diabetes can be obtained in such mice following implantation of beta cell precursors derived from human ES cells. Based on the animal experiments and the current clinical procedure of islet transplantation, we can conclude that if a pure and well characterized population of beta cells can be made in culture, it is likely that they will prove effective for diabetes treatment. This is why diabetes is always listed as one of the front runners for pluripotent stem cell therapy.

However we can also predict that progress will be slow. Despite all the problems arising from diabetes, the current treatments for the disease are quite good, and promise a lifespan approaching normal to those patients who can achieve good blood glucose control. Any new treatment will not receive regulatory approval unless it is at least as good, and preferably better, than the existing treatments, and any risks from a new treatment need to be proportionate to the situation. If a disease causes 100 per cent mortality in 6 months, then a treatment with an estimated 10 per cent mortality from side effects might be acceptable. But this is not the case for diabetes, and any new treatment will have to meet a very high threshold of safety.

Moreover, pluripotent cell-derived beta cells are not the only way of tackling the problem. Another possibility is to obtain beta cells from animals, especially pigs. This presents formidable problems of overcoming graft rejection, but some immunologists are confident that they can be solved. There is also the engineering approach. Insulin pumps are already very sophisticated, and it is possible to build so called 'closed loop' devices which automatically monitor the blood glucose and regularly deliver the amount of insulin required.

For all these reasons, the future is hard to discern. If the technical problems of making beta cells can be overcome, if safety can be guaranteed, and if an acceptable solution can be found to the problem of autoimmunity, then pluripotent cell-derived grafts may become commonplace for type 1 diabetes. In fact they may even be used to treat the more numerous type 2 diabetic population. But if these objectives are not reached, or if competing therapies prove more effective, then despite all the hopes and publicity, this technology will remain in the sidelines as a niche solution rather than a revolutionary cure.

Parkinson's disease

Parkinson's disease is a progressive disorder affecting movement. It is relatively common, with a lifetime risk of 1–2 per cent being the second most common neurodegenerative condition after Alzheimer's disease. The average age of onset is 60 and the disease is characterized by rigidity, tremor, slow movements, and, in extreme cases, inability to move. Although Parkinson's disease primarily affects movement it can also affect language and cognitive ability. The movement problems arise from a decreased stimulation of the motor region of the cerebral cortex. This arises from a loss of a particular class of neuron in the brainstem, those which use dopamine as a neurotransmitter. Neurons communicate with one another by releasing small amounts of neurotransmitter substances at the synapses which connect the projections of one neuron with the cell body of another. The secretion of the neurotransmitter will activate or inhibit the receiving cell, and this is the basis of all neuronal circuits. The loss of dopamine-producing neurons means that the motor cortex does not receive the correct input signals and so the motor system nerves are unable to control movement and coordination properly. Dopamine is produced by a specific class of neurons, called dopaminergic neurons, in the brainstem, especially in the substantia nigra region. In Parkinson's disease, the degenerating neurons die and in the course of their degeneration they display structures called Lewy bodies. Parkinson's disease patients have lost 80 per cent or more of their dopamine-producing cells by the time symptoms appear.

Dopamine is normally made from a precursor substance called L-DOPA. From the 1960s, L-DOPA, along with ancillary drugs, has been administered to treat Parkinson's disease. It does have beneficial effects but these are accompanied by some problems with uncontrolled movements, and side effects such as mood disorders and sleep disturbances. As the disease progresses, the freezing to immobility and the loss of cognitive functions may cease to respond to L-DOPA.

Foetal midbrain grafts

At first sight Parkinson's disease appears an ideal candidate for cell replacement therapy because the disease is based on the loss of one specific cell type, the dopaminergic neuron. However, there are a number of problems with neuronal cell therapy that make the problem intrinsically more difficult than for diabetes. An endocrine cell, such as the pancreatic beta cell, exerts its function by secreting a substance into the blood stream, insulin in the case of beta cells. Within limits, endocrine cells are able to function anywhere in the body where they have access to a blood supply, and they do not necessarily need specific connections to nerves or other cell types. By contrast, neurons function by emitting neurotransmitters that stimulate or inhibit the action of other neurons. This means that they must be in the right place and have the right inputs and outputs connecting them to other neurons, a situation posing considerable challenges for cell delivery.

Nonetheless, there is a long history of cell therapy for Parkinson's disease. Initiated in the 1980s by Dr Olle Lindvall in Lund, Sweden, there have been a number of small trials involving the implantation of grafts of tissue from the midbrains of 6–9-week-old human foetuses into the brain region to which the dopaminergic neurons normally project (called the striatum). But unlike the islet transplantation for diabetes there is some dispute about whether this type of cell therapy is actually effective. On the positive side, some patients have experienced substantial improvement of their disease. Also it has been clearly established from post-mortem studies of patients who subsequently died that the grafts do generate abundant dopaminergic neurons and that these can survive long term and connect to other neurons. On the other hand, randomized trials have shown no benefit of the human midbrain graft over placebo in terms of disease progression, raising the suspicion that the positive effects are due to chance and not to the treatment. Some patients have developed complications involving uncontrolled movement (also experienced as a complication of L-DOPA treatment). Also, rather surprisingly, some post mortem studies after many years have indicated that cells of the graft can develop the Lewy bodies, indicating that the disease can spread from the host to the graft cells. The mechanism for this spread is unknown. So, although some practitioners are convinced that the randomized trials were faulty and that foetal midbrain grafts are effective, it is not absolutely certain that a ready supply of dopaminergic neurons would transform the treatment of Parkinson's disease.

Human pluripotent stem cells are considered a suitable source of dopaminergic neurons because the supply of early human foetuses as graft donors is limited. Also, since they are obtained as a result of elective abortion, there are inevitably some ethical issues. Abortion is opposed by more people than oppose the use of human preimplantation embryos to grow embryonic stem cells. So there is a significant body of opinion who consider it is all right to use human embryonic stem cells but not all right to use human foetal grafts. Accordingly, Parkinson's disease is usually listed among the priority diseases for treatment with human ES cells, or more recently, with iPS cells.

Many labs have developed methods for the differentiation of human pluripotent stem cells into dopaminergic neurons. The published methods differ to some extent between laboratories, but they generally involve a stage of embryoid body formation, followed by a neural induction process and the cultivation of cells resembling neural stem cells as aggregates in suspension. These are then treated with inducing factors known to induce formation of the midbrain in normal embryonic development and differentiated as a monolayer culture.

The standard animal model for cell therapy experiments is a rat which has been injected on one side of the brain with 6-hydroxydopamine, a substance resembling dopamine that causes destruction of the dopaminergic neurons. This one-sided lesion leads to various asymmetrical behaviour patterns. The putative therapeutic cells are injected into the affected side of the brain, usually into the striatum which is the region to which dopaminergic neurons normally connect, and the efficacy is assessed by a battery of behavioural tests. After a suitable period of time the rats are sacrificed and their brains are analysed for the persistence and differentiation state of the transplanted cells. Experiments of this sort do indicate effectiveness of the treatment in terms of improving behaviours. They also indicate long-term persistence of the grafted cells, including many dopaminergic neurons, and there are usually no teratomas formed.

Alternative treatments for Parkinson's disease

The cell production technology and the results of animal experiments look favourable when contemplating the use of dopaminergic neurons derived from human pluripotent stem cells for the treatment of Parkinson's disease. But apparently revolutionary medical innovations are often tripped up by another consideration. A new treatment has to be better than existing treatments and has to have at least no more risk of adverse side effects than existing treatments. There are several treatments for Parkinson's disease in current use. Although administration of L-DOPA and associated drug therapy has limitations, it is effective to some degree in the early stages of the disease. Moreover, a completely new type of treatment has also been under development for some years. 'Deep brain stimulation' is a technique that involves the implantation of a device called a brain pacemaker which sends electrical impulses to specific regions. It directly changes brain activity in a controlled manner and its effects are reversible. It has been under investigation for about 20 years for treatment of various neurological conditions. The mechanism of action of deep brain stimulation is unclear, but it does seem to work and the FDA approved it as a treatment for Parkinson's disease in 2002. Given the equivocal status of the clinical trials of the human foetal midbrain grafts, it is not certain that doperminergic cell grafts would perform better than deep brain stimulation.

There are also some potential safety concerns with grafts derived from pluripotent cells. Although Parkinson's disease is a distressing condition that causes much suffering, it does not greatly reduce lifespan, so the period of time available for the development of complications is quite long, perhaps about 20 years on average. For any cell therapy derived from pluripotent stem cells there is always an issue about the possibility of persistence of a few pluripotent cells in the graft which might give rise to teratomas. Although the animal experiments indicate that the teratoma risk is very low, the long survival time of Parkinson's patients does make even a very small cancer risk seem significant. Moreover, the differentiation protocols for pluripotent stem cells never produce 100 per cent of the desired cell type. Even if all the pluripotent cells are gone there will certainly be other types of neuron and *glial cell* present and these may generate unwanted effects. For example, the uncontrolled movement problems seen in some of the foetal midbrain graft recipients has been ascribed to the presence of other types of neuron which make inappropriate connections.

So, although the technological problems of cell therapy for Parkinson's disease are largely solved, the risk-benefit ratio in favour of this treatment as compared to others is not overwhelming at present.

Heart disease

The heart is an organ of which people are generally much more conscious than they are of their pancreatic islets or of the motor circuits in their brainstem. Everyone knows that the heart pumps the blood round the body and that if the heart stops you will die in a few minutes from lack of oxygen to the brain. In order to keep beating every minute of every day for 80 years or so, the heart has to be very robust. Most of the heart is composed of a specialized type of muscle cell called the cardiomyocyte. Unlike skeletal muscle, cardiomyocytes contract spontaneously but in the heart they do so in a concerted manner in response to electrical signals from special pacemaker regions. The pacemakers, in turn, are controlled by various hormonal, metabolic and neuronal inputs that affect the heart rate. The continued contraction of the cardiomyocytes of the heart depends on an abundant blood supply that brings oxygen and nutrients to the cells via the coronary arteries. The most common form of sudden death in modern developed societies is the heart attack (myocardial infarction), in which one or more coronary arteries becomes blocked and the sector of heart muscle that it normally supplies becomes deprived of oxygen. Unless the blockage resolves spontaneously, or as a result of emergency treatment, the affected region of heart muscle will die in about one hour. If the damage is extensive enough to abolish most cardiac function, the individual will also die. If the affected region is more limited, the patient will survive but with a permanently damaged heart. There is some dispute about whether cardiomyocytes are normally replaced during life, but if they are replaced it is at a very slow rate, and there does not appear to be any significant regeneration in areas of damage. Instead the damaged areas fill up with cardiac fibroblasts which secrete extracellular matrix material, and so the region of dead muscle will become replaced by a scar. This has mechanical integrity but is inactive when it comes to contraction, and so the heart function is correspondingly diminished. If the heart is subjected to a lot of excess stress then surviving cardiomyocytes will become enlarged and less mechanically competent, leading eventually to heart failure. The loss of function due to heart attacks is the most common cause for this, although there are other causes such as high blood pressure or valve disease. By definition, heart failure means an inability to supply the rest of the body with enough blood. It leads to numerous problems and is likely to prove fatal in a shorter or longer period, depending on severity.

Cell therapy of the heart

Although there is considerable ongoing activity which could be described as cell therapy of the heart, this is quite different in character from the islet cell grafts used to treat diabetes and the foetal midbrain grafts used to treat Parkinson's disease. These two examples are both of cell replacement with a specific type of differentiated cell that has been lost from the patient. By contrast, there is no clinical cell therapy of the heart based on the introduction of new cardiomyocytes. There is a great deal of aspirational stem cell therapy based on the injection of various cell populations drawn from elsewhere in the same patient, and this will be discussed later. Whether it is effective or not, this type of therapy certainly does not replace cardiac muscle.

However, in the belief that grafts of healthy heart muscle cells could help reverse the effects of heart attacks, many groups are developing methods to make cardiomyocytes from human pluripotent stem cells, and testing protocols for cell replacement in animal experiments. As in the other cases, the differentiation protocols depend on guiding the pluripotent cell through a series of developmental steps through exposure to a sequence of inducing factors. First mesoderm needs to be formed, then anterior type mesoderm, then cardiac progenitor cells, and finally the cardiomyocytes themselves. In this context it should be noted that even if the proposed cell therapy never actually happens, the production of human cardiomyocytes is in itself a very valuable objective. These cells are important for testing new drugs, not just those intended to act on the heart, but also all other drugs in line for human clinical trials because adverse side effects on the heart are so common. Animal experiments have some value in predicting these effects, but more reliable results may be obtained with human cells, and for obvious reasons live human heart muscle cells are generally not available.

The animal models for cell therapy of the heart usually involve a coronary artery ligation to cause damage in a specific sector of the heart muscle. Then the cells are injected into the affected part, and the heart function is studied with a battery of physiological tests. Eventually the animal is sacrificed and the presence, differentiation state and spatial distribution of the graft-derived cells can be established. There are experimental difficulties associated with the fact that rat or mouse hearts beat at several hundred beats per minute, rather than the 60–80 of the human heart. Experiments with large animal hosts such as pigs are considered more reliable although they are very expensive and need complex facilities to carry them out. In general, published results of animal experiments show persistence of the grafted cells and some improvement in cardiac function. However, the interpretation is complicated by the fact that the grafts generally contain progenitor cells for blood vessels as well as cardiomyocytes themselves. Indeed, it is widely believed that the normal cardiac progenitor cell in embryonic development is a multipotent cell that can form either the endothelial lining of blood vessels, or smooth muscle which is found in the outer layer of arteries, or the cardiomyocytes themselves. More blood vessels may improve blood supply to the damaged region and improve the function of residual heart muscle. While this would still represent a positive result, it is not genuine cell replacement therapy of the cardiac muscle.

Spinal repair

The first clinical trial of cells derived from human ES cells was approved by the FDA in 2009. Surprisingly, this was not for one of the perceived front runners, like diabetes, or Parkinson's disease or heart disease, but rather for the use of remyelinating cells to treat spinal trauma.

Spinal trauma is a rather difficult condition for cell therapy. The paralysis and loss of sensation below the level of the injury is typically caused by damage to nerve fibre tracts descending (motor) or ascending (sensory) the spinal cord. The cell bodies from which the fibres of the tracts originate are likely to be at remote locations and may still be alive. Spinal trauma also causes local cell death of both neurons and the glial cells that surround them. Also, damaged areas are transformed into scar tissue. This is somewhat different from familiar soft tissue scarring, being due to the proliferation of a class of glial cell called astrocytes, but it has the effect of inhibiting any regrowth of nerve fibres through the region of the scar. In addition, some fibres that are not otherwise damaged may be subject to loss of their insulating myelin sheaths, and this loss prevents efficient electrical signal transmission by the fibres.

The rationale for the first trial of ES-derived cell therapy of spinal injury is to put in healthy *oligodendrocytes*, which are the class of glial cell normally responsible producing the myelin sheaths of axons, in the hope that some of the spared fibres will be remyelinated and will then recover functional activity. The trial is being conducted by the biotech company Geron Corp., based

Chapter 5
Tissue-specific stem cells

e stem cells found in the *postnatal* body (the body after birth) the tissue-specific stem cells responsible for tissue renewal, or repair following damage. They share with pluripotent stem s the definition of stem cells provided in Chapter 1, namely bility to reproduce themselves and to generate differentiated geny cells. But they share few molecular characteristics with or iPS cells, such as expression of specific transcription factors ll surface molecules. As mentioned earlier, the definition of cell is based on biological behaviour rather than on intrinsic acteristics.

turnover in the body

der to understand about tissue-specific stem cells, it is useful nk about the body and how it grows, persists, and renews The process of early embryonic development generates a er of zones of cells each committed to form specific body Usually more than one such zone contributes the cells for rgan. For example, the small intestine contains an inner of epithelial cells which derives from a segment of the erm of the embryo. Around this are several layers of smooth , connective tissue, and blood vessels that derive from a c part of the mesoderm. In addition, the intestine contains erve cells, derived from the ectoderm of the embryo, and in California, and uses oligodendrocytes made from human ES cells. Extensive safety testing in animals persuaded the FDA that there was a minimal risk of teratoma formation. Even so, progress has been cautious. The trial was suspended for 18 months at an early stage because of worries about possible tumour risks, and restarted in 2010. As a phase 1 trial, its aim is primarily to establish safety of the treatment. Serious investigation of efficacy will need to await a larger phase 2 trial.

Retinal degeneration

In the eye, in the centre of the retina, lies a small (5 mm diameter) pigmented area called the macula. It contains the highest density of cone-type photoreceptors (light receptors) in the retina and is responsible for the high level of detail and discrimination in the central part of the visual field. It is quite common for this part of the retina to deteriorate with age, with loss of photoreceptors, and about 10 per cent of individuals over 65 have some degree of age-related macular degeneration (ARMD). In severe cases this can lead to loss of central vision, which prevents reading, recognition of faces, and other tasks requiring high visual acuity. Although some peripheral vision remains, many patients with severe ARMD are legally classified as blind.

There are two main forms of ARMD. The 'dry' form is associated with the appearance of debris in the region and is thought to be due to a defect in the retinal pigment epithelium (RPE), which is a layer of pigmented cells lying beneath the photoreceptors. The 'wet' form is a type of overgrowth of blood vessels from the eye capsule (the choroid) into the subretinal space. There is no treatment for the dry form. The wet form can be treated by laser ablation of the new blood vessels and/or by injection of specific antibodies that antagonize blood vessel growth.

As with the other conditions discussed in this chapter, there has been a background of transplantation therapy, both in human various immune cells derived from the haematopoietic (blood-forming) system, also part of the mesoderm.

Within our bodies there is a continuous cycle of replacement, so the molecules of which we are composed are not the same from year to year, even though they occupy similar positions. But there is an important difference between metabolic turnover, which leads to the renewal of substances within cells, and cell turnover, which involves the birth of new cells and the death of old ones. Most substances turn over in most tissues, the exceptions being DNA in non-dividing cells and certain extracellular proteins such as those in the lens of the eye, which can literally last a lifetime. But by no means all cells become renewed by cell division, and it is the presence of cellular renewal that is relevant to determining the existence and identity of tissue-specific stem cells.

The division of a cell into two daughter cells is the end point of a process called the cell cycle. One of the essential events that occurs in the cell cycle is DNA replication: the DNA molecular double helix separates into two strands and a complementary copy is assembled on each to generate two double helices, each identical to the original. This DNA is packaged into chromosomes. Shortly before cell division takes place the chromosomes 'condense' into short fat bodies, which are arranged at the midpoint of a spindle composed of structures called microtubules. Then the cell physically divides into two. In the course of this, the microtubules draw the two chromosome sets apart and they become packaged in separate nuclei. Then the cytoplasm constricts to separate the two daughter cells. The whole process of cell division is called *mitosis*.

The father of the study of cell renewal was Charles Philippe Leblond who worked at McGill University in Montreal, and was an expert in histology: the study of human or animal tissues at the cellular level, viewed as thin sections down the microscope. After the Second World War, Leblond made a systematic study of all the tissues in the body in terms of whether they contained dividing cells and where those cells were located. He classified tissues into three fundamental types: post-mitotic, expanding and renewal. The *post-mitotic* cells are formed during embryogenesis or soon after birth, and never divide thereafter. The main examples are neurons and muscle fibres. We now know that there is some ability to generate new neurons and muscle fibres in adult life, but once generated these cell types are indeed post-mitotic and never divide again. The 'expanding' type refers to expansion of numbers during normal juvenile growth with cessation of cell divisions once full size is reached. This group comprises most of the tissues of the body, including most connective tissues and epithelial organs (e.g. liver, kidney, adrenal, thyroid). Many of these tissues do have some regenerative capacity but in the absence of damage they are mostly non-dividing in adult life. The most interesting of Leblond's categories was the *renewal tissues*. These are tissues whose cells are continuously being replaced. Each possesses a population of stem cells, which persist for the life of the organism, and which continuously generate new cells to repopulate the tissue. In the non-growing adult the generation of new cells by division is precisely matched to the removal of old cells by cell death.

For example, the epithelial lining of the small intestine is a renewal tissue. The small intestine contains an array of small finger-like villi projecting into the intestinal interior (lumen), which provide a large area for the absorption of digested food (Figure 11). In between the villi are tiny pits called crypts of Lieberkühn (named after a German physician, John Lieberkühn, who first described them in 1745). All of the dividing cells in the intestinal epithelium are located in the lower part of the crypts. Cells are continuously produced in the crypts, then move up the villi and perform their functions for a few days before they die and are shed from the villus tips (Figure 12). Subsequent study has shown that each crypt contains about six stem cells. They divide to produce progenitors called transit amplifying cells which each divide several more times before becoming post-mitotic.

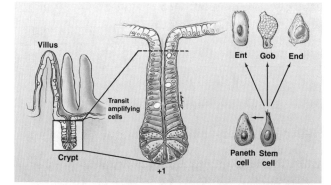

11. **Intestinal stem cells. These are found at the base of the intestinal crypts. Above them are dividing transit amplifying cells. The intestinal stem cells give rise to all four types of differentiated cell found in the intestinal epithelium. Cells are continuously produced in the crypts, migrate to the villi, die, and are shed into the intestine**

This means that the majority of dividing cells in the crypt are not the stem cells themselves but are actually transit amplifying cells. Each stem cell generates progeny that differentiate into all of the cell types in the intestinal epithelium, comprising the regular absorptive cells, the goblet cells that secrete mucus, the Paneth cells that defend against microorganisms, and the entero-endocrine cells that release a variety of hormones to control the functions of the intestine.

Some other renewal tissues described by Leblond are the other epithelia of the gut (stomach, large intestine), the epidermis of the skin, the testes, and the haematopoietic system. As we saw in Chapter 1, the epidermis continuously generates cells from its basal layer. Only a small proportion of the basal layer cells are stem cells, most are transit amplifying cells. They cease dividing while still in the basal layer and migrate out into the upper layers then progressively differentiate into the mature epidermal cells of the skin. Eventually they die and are shed from the skin surface.

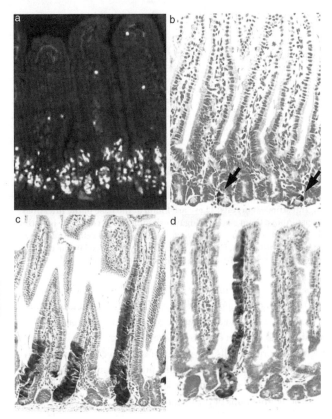

12. Microscope sections of the intestinal epithelium of a mouse.
(a) The cells of the intestinal crypts have been labelled with a pulse
of bromodeoxyuridine (BrdU), which is incorporated into DNA of
dividing cells and visualized with a specific antibody. (b)–(d) The
progeny of intestinal stem cells following introduction of a genetic
label into a few individual stem cells, 1 day after labelling (b), 5 days
after labelling (c) and 60 days after labelling (d). Because this is a
genetic label it remains in the stem cell and all its progeny, which form
a file of cells leading up to the villus tip

In the testes, sperm are continuously generated from dividing cells called spermatogonia which occupy the outer layer of cells in the seminiferous tubules. A small proportion of these are stem cells and the majority are transit amplifying cells, dividing a few times before differentiating into spermacytes. Spermacytes then divide to form four spermatids, each of which differentiates into a mature sperm.

The haematopoietic system

The *haematopoietic* system is located in the bone marrow of adults and generates all the cells of the blood and immune system. The haematopoietic stem cells (HSCs) are only a small fraction of all the cells in the bone marrow but, over time, they produce all of the other cells. This system has been studied intensively because of the importance of *haematopoietic stem cell transplantation* (*HSCT*), comprising bone marrow transplantation and allied techniques.

The HSCs can be isolated using a fluorescence-activated cell sorting (FACS) machine which can purify single cells from a complex mixture based on the specific combination of cell surface molecules that the cells possess. The methods for doing this were perfected by Irving Weissman of Stanford University. HSCs are particularly sensitive to radiation, so if an animal receives enough radiation all them are destroyed, and without treatment the animal will die of marrow failure. However, it is possible to rescue irradiated animals by injecting HSCs into their bloodstream (either as pure stem cells or as crude bone marrow). The injected HSCs can locate niches in the bone marrow and, over time, reconstitute the entire blood and immune system such that all of these cell populations in the host are derived from the HSCs of the graft. There is no routine method to expand haematopoietic stem cells *in vitro*, so both animal experiments and human transplantations are carried out using freshly isolated cells.

It has been shown that the HSCs are a small fraction of the dividing cells and are located in the marrow either in association with bone cells or with the blood vessels. They divide to renew themselves and also to generate transit amplifying cell types called the common lymphocyte progenitor and the common myeloid progenitor. As their names suggest, these cell types can form, respectively, all of the types of lymphocyte (T cells, B cells, and NK cells), and all the types of myeloid cells ('myeloid' simply means 'of the marrow') which include red blood cells, macrophages, granulocytes, and blood platelets.

Like other tissue-specific stem cells, HSCs can form only the differentiated cell types of their own tissue, in this case the blood and immune system. Claims that they could contribute to a wide range of other tissue types were made around the turn of the 21st century and many labs conducted experiments in which genetically labelled HSCs were transferred (grafted) from one animal to another. Normally the graft gives rise to all cells of the blood and immune system which will therefore be labelled with the genetic marker characteristic of the graft. The issue was: do such grafts also contribute cells to other structures such as the heart, brain, pancreas or liver? After many studies, the consensus is that they do not. There are low level contributions of cells to many tissues, but these are ascribed either to trapping of cells within other organs, fusion of donor and host cells, or uptake of marker genes by host cells. There may be a very low level of genuine differentiation into other tissue types, especially if the host animal has been heavily irradiated, but it is insufficient to lead to real functional regeneration and is certainly too low to be useful in clinical cell therapy.

Studying cell turnover

A number of methods for studying the renewal of cells (cell turnover) have been introduced since the original studies of Charles Leblond. Originally Leblond just looked for dividing cells under the microscope. These can be identified as *mitotic figures*, which are the patterns adopted by the chromosomes as they segregate into the two daughter cells. However, the process of mitosis is over quite quickly, so the proportion of mitotic figures visible in a section is much lower than the actual proportion of cells undergoing cell cycles. Nowadays dividing cells are more readily identified by tagging ('staining') using antibodies that recognize specific proteins whose levels are elevated in dividing cells. One such antibody commonly used in histology labs is called Ki-67 after the town of Kiel in Germany where it was discovered. Another approach, useful for animal experiments, is to administer a precursor substance which becomes incorporated into new DNA molecules as they are synthesized and can subsequently be identified by staining with a specific antibody. The most commonly used is *bromodeoxyuridine (BrdU)*. This is chemically similar to thymidine, which is one of the four nucleotides found in DNA. If BrdU is injected into an animal, some of its molecules become incorporated in place of thymidine into the DNA of all the cells that are synthesizing DNA at that time. The beauty of using BrdU is that after a while it is all used up and then subsequent DNA synthesis will use the endogenous supply of thymidine as normal. If the animal is sacrificed and its tissues examined down the microscope, the only cells labelled with BrdU will be those that were in cycle shortly after the BrdU injection (Figure 12a). If they subsequently stop dividing they will retain the BrdU because nuclear DNA persists forever and does not turn over by metabolism. But if they continue to divide they will lose the BrdU because it will become diluted by a factor of two every time the DNA is copied. After about five cycles it will have been diluted by 32 times and will probably no longer be detectable. This makes BrdU a very useful tool and with it scientists have characterized the cell turnover kinetics of all the renewal tissues, identified specific dividing populations in regenerating tissues, and identified the time of final division (the so-called 'birthday') of neurons and other post-mitotic cell types.

Do non-renewal tissues contain stem cells?

It is often implied that all tissue types contain their own tissue-specific stem cells. However this is not correct. Only the renewal tissues really have stem cells in the sense of a special population of cells that reproduce themselves and that continue to generate differentiated progeny over the lifetime of the organism. But there are some types of tissue regeneration behaviour involving special cell populations that may also be considered stem cells. For example, the fibres of skeletal muscle are post-mitotic. They are formed in embryonic development by fusion of muscle progenitor cells which arise in the mesoderm of the embryo. But skeletal muscle also contains a population of *muscle satellite cells* that are small undifferentiated cells found beneath the outer sheath surrounding each muscle fibre. Muscle satellite cells are normally quiescent, but following damage to the muscle they can become activated and will proliferate. In the course of this proliferation they both renew themselves and generate muscle progenitor cells which can either fuse with existing fibres or fuse with each other to generate new fibres. It appears that the self-renewal is not perfect so the supply of muscle satellite cells does decline with age, and in this sense they do not quite fulfill the requirements for being true stem cells. However, they are normally considered as a type of stem cell and there has been much interest in isolating muscle satellite cells and expanding them in culture with a view to transplantation therapy of diseases where muscle fibres are destroyed, most notably Duchenne muscular dystrophy.

An organ that has considerable regenerative capacity is the liver. This is probably because the liver is the first destination for absorbed foodstuffs from the intestine, conveyed through the hepatic portal vein, and these always contain a certain level of toxins. Most plants contain naturally occurring toxins to defend against insect attack and these will kill a few liver cells in the normal course of events. The steady state level of cell division in the liver is low, but if a dose of toxin causes some cells to die they are replaced very quickly. They are also replaced following surgical removal of part of the liver, a fact possibly known to the ancient Greeks when they composed the legend of Prometheus, who stole fire from the heavens and was punished by being chained to a rock and having his liver eaten each day by an eagle. However, this type of liver regeneration does not occur from a special population of stem cells. It is the normal functional liver cells making up the main mass of the liver tissue, the hepatocytes, that divide to produce the new hepatocytes. There are also some cells called *oval cells*, located in the bile duct regions of the liver, which are thought to be capable of becoming either hepatocytes or bile duct epithelial cells. They can be provoked to regenerate parts of the liver in certain types of animal experiment where division of the regular hepatocytes is suppressed by drug treatment. Whether the oval cells are themselves stem cells, or are derived from a population of stem cells, is currently unclear.

In the pancreas there has been a strong desire to find some sort of stem cell population that might be exploited to generate beta cells for transplantation therapy. But the pancreas resembles the liver in that there is normally very little cell division. If the pancreas is damaged by administering a toxin, the exocrine cells, responsible for the production of digestive juices, do regenerate quite fast, but the endocrine cells, which produce the insulin and other hormones, do not. Animal experiments using BrdU labelling show that regeneration of exocrine cells occurs not from a specialized pool of stem cells but from other surviving exocrine cells in the vicinity. Furthermore, the normal increase in the number of beta cells that occurs during growth of the animal, or during pregnancy, occurs not from stem cells but from pre-existing beta cells.

Do brain or heart cells turn over at all?

Just as there is a strong desire to believe in stem cells in the pancreas, this desire is perhaps even stronger for the brain and the heart, both vital organs that are vulnerable to debilitating and lethal diseases involving loss of viable cells. In both cases, the principal functional cell type—the neurons in the brain and the cardiomyocytes in the heart—belong to the post-mitotic category of Leblond. These cell types certainly are post-mitotic, but the issue of whether any new, additional, neurons or cardiomyocytes are formed during adult life has been a controversial one.

In the case of the heart, BrdU does not generally label any cardiomyocytes and this indicates that any new heart muscle formation must occur very slowly, if at all. On the other hand, the formation of new neurons has been clearly demonstrated in two locations in the brain: the lining of the lateral ventricles, from where new neurons migrate to the olfactory bulb, involved in the sense of smell; and the dentate gyrus, which is a part of the hippocampus, a structure concerned with learning and memory. Both these regions contain *neural stem cells* which seem similar to the neural progenitor cells present in the foetus and are capable of generating both neurons, which are the functional cells of the brain, and *glial cells*, which perform structural and supportive functions. These areas of cell renewal have been demonstrated by BrdU labelling in animal experiments and also in human cancer patients who were injected with BrdU for diagnostic purposes and gave permission for post-mortem examination of their brains. Animal studies on other parts of the brain, such as the all important cerebral cortex, have not yielded convincing evidence of any cell renewal either during normal life or following tissue damage. This is in contrast to the situation in lower vertebrates (amphibians and fish) which do show cell turnover and significant ability to regenerate most parts of the brain following damage.

Even BrdU labelling is not sensitive enough to detect very slow rates of cell turnover and there is always the possibility that cortical neurons or cardiomyocytes of the heart are renewed to some degree, either normally or following damage, but at too slow a rate to be observable. Fortunately there is now a method of observing slow cell turnover, and in humans rather than lab animals.

This was introduced by Jonas Frisén and Kirsty Spalding at the Karolinska Institute in Sweden. It relies on the fact that in the 1950s and early 1960s the governments of the USA, USSR, Britain and France carried out an extensive radiolabelling experiment on the entire population of the world without their consent. These countries carried out numerous atmospheric tests of nuclear weapons from the late 1940s up to the Limited Test Ban Treaty of 1963, after which all nuclear tests were conducted underground. Many radioisotopes were released into the atmosphere from the tests, and were rapidly carried around the world by the winds. The *isotope* that is used for cell turnover analysis is carbon-14 (^{14}C). This is familiar for its use in archaeology, where dating studies make use of its radioactive decay. However, the use of ^{14}C for cell turnover studies is not based on radioactive decay but on the much faster loss from the atmosphere due to incorporation into the sea and into geological sediments. Since plants absorb carbon dioxide (CO_2) from the atmosphere for photosynthesis, and animals (including ourselves) eat these plants, our intake of carbon in a given year has an isotopic composition very similar to that of the atmosphere at the same time. Determination of ^{14}C abundance in tree rings, whose date of formation is precisely known, shows a rapid increase from 1955 to 1963 due to the nuclear tests, followed by a slower exponential decline, due to the removal of ^{14}C from the air (Figure 13). Radioactive decay of ^{14}C is negligible over this period, and does not affect the results, since the radioactive half life is 5730 years.

Imagine that a person was born in 1965, and died in 2010. For substances for which there is no metabolic turnover, including the DNA of cells that have not divided since birth, ^{14}C that was incorporated during foetal life in 1964–65 will still be present at the same level of abundance. However, molecules that have been synthesized more recently, including the DNA of cells that have divided since birth, will have a lower ^{14}C abundance corresponding to the atmospheric level in their year of synthesis. In the case of

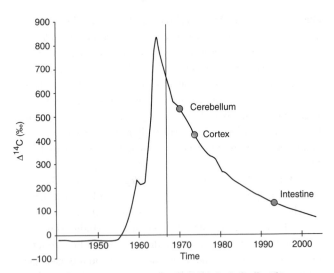

13. The ^{14}C method for determining the 'birthday' of cells. The curve shows the atmospheric abundance of ^{14}C between 1940 and 2000. The individual denoted by the vertical line was born in 1967 and died in the early 2000s. The ^{14}C abundance in DNA from the intestinal epithelium indicates a low cell age relative to the age of the individual, while that from the brain (cerebellum and cortex) indicates a greater age. This figure includes all cell types from the brain and intestine, better precision is achieved by examining specific cell types

renewal tissues like the blood or epidermis, where cells turn over rapidly, the DNA will have a ^{14}C content approximating to 2010.

Thus, in principle, it is a simple matter to determine the age of any cell, by measuring the ^{14}C abundance of its DNA. In reality it is a very complex process, as the mass spectrometer required to make the measurements of ^{14}C/^{12}C ratio is very sophisticated, and it is necessary to purify cell nuclei of the specific cell type under examination before extracting the DNA. This last point is particularly important. All tissues contain many different cell types, and it is no good just sampling whole brain or heart, because they contain many blood vessels, immune cells, fibroblasts and so on, that are not the cells of interest.

The Karolinska lab has surmounted all the technical problems and published studies on several tissues. In the brain they isolated nuclei from neurons of the cerebral cortex, the part of the brain responsible for higher functions. The results indicate that there is no renewal of neurons whatsoever. All the cerebral cortex neurons were formed during foetal life and they are not added to after this time, at least not enough to alter the isotope ratio. So, despite the fact that certain other areas of the brain do show some cell renewal, for the all important cerebral cortex we really are living with our lifetime supply of cells from the time of birth. This is one reason why neurodegenerative diseases are so devastating: when central neurons die, nothing can bring them back.

A similar study on the heart suggests that there might be a little cell turnover, but it is less than 1 per cent per year, declining with age. In other words, the entire heart would not be replaced within a lifetime of 80 years, so if there is any renewal at all it occurs at a very modest rate.

Neurospheres

Despite the restriction of stem cells to just two regions of the mammalian central nervous system (CNS)—the lateral ventricle linings and the dentate gyrus—it is possible to grow neural stem cells in tissue culture from a much wider range of brain regions than this. The stem cells grow in a form called *neurospheres*. These are clumps of cells, up to 0.3 mm wide, that grow in suspension culture in a medium containing two specific growth factors (EGF and FGF). The cultures can be initiated from any part of the foetal CNS and often from parts of the adult CNS as well, even regions not thought to undergo continuous renewal. Neurospheres are thought each to contain a few neural stem cells, which are capable of self-renewal, plus a certain number of transit amplifying cells, that have finite division potential. When neurospheres are plated on an adhesive surface in the presence of serum, they will differentiate and form the three cell types normally generated by neuronal stem cells, which are neurons, and two types of glial cells: *astrocytes* and *oligodendrocytes*. If neurospheres are dissociated into single cells, a few per cent of these cells can establish new neurospheres, with similar properties to the original. Repeated cycles of dissociation and growth can provide substantial expansion capacity.

The phenomenon of neurospheres is an example of the fact that cells may behave in a different manner in tissue culture from *in vivo*. Neurospheres have created enormous interest because, unlike haematopoietic stem cells, they are expandable *in vitro*, and because there is a hope that they might be used for cell therapy of the very intractable neurodegenerative diseases involving widespread neuronal death.

Pluripotent stem cells in the adult

One good reason for preferring not to use the term 'adult stem cell' is that it has become associated with the idea of ES cell-like cells in the adult organism. It is very unlikely that these exist. The science of developmental biology would predict that there are no pluripotent cells left in an organism after birth because all will have been programmed to follow some specific developmental pathway during the first steps of embryogenesis. However the example of neurospheres shows that it may be possible to use the tissue culture environment to obtain useful cell populations whose properties are somewhat altered from those *in vivo*.

Apart from neurospheres, there are certain other types of multipotent (although not pluripotent) cell that are normally quiescent but can be roused to proliferative activity in culture. If we take the ability to repeat studies in different labs as the main index of credibility, then the front runners are the *mesenchymal stem cells* found in the bone marrow. These are also known as 'marrow stromal cells' both fortunately abbreviating to MSC. These cells certainly exist. They were first discovered by Alexander Friedenstein, working at a research institute in Moscow in the 1970s and have been studied by many labs ever since. They are quite different from the haematopoietic stem cells (HSCs) also found in the bone marrow. Unlike the HSCs they can be cultured *in vitro* and will grow as an adherent cell layer on plastic. They can become several differentiated cell types, especially bone, cartilage, fat or smooth muscle cells, when cultured in suitable media. Similar cells can be isolated from other tissues especially umbilical cord blood and adipose (fat) tissue, which is for obvious reasons viewed as a potentially limitless source of human material. There remain real uncertainties about the functions of MSCs *in vivo*. It is not known exactly where in the tissues they are located, whether they actually function as stem cells (i.e. showing lifelong self-renewal and generation of differentiated progeny), and what range of cell types they actually generate in the body. It is quite likely that they are the connective tissue equivalents of the muscle satellite cells or oval cells in the liver, functioning as an occasional source of cells for regeneration rather than for continuous tissue turnover.

There are also literally dozens of types of postnatal 'stem cell' that have been described from time to time as showing fully pluripotent behaviour resembling that of ES cells. However none has so far proved reproducible between labs and so none has become generally accepted as a real entity by the scientific community. This is a very important practical issue because the dubious 'adult stem cells' of this sort are often those advertised by companies and clinics as the vehicle for their aspirational stem cell cures. The irreproducibility of such reports has been attributed to chance events such as the occurrence of mutations in the cells, combined with the effects of selection of fast growing sub-populations during growth in culture.

The contrast with iPS cells is very instructive. When iPS cells were first discovered, or perhaps one should say created, by Shinya Yamanaka in 2006, there was initially some scepticism from the scientific community. But the work was rapidly repeated and extended in other labs. Now iPS cells are made by hundreds of labs around the world, using a variety of methods, and there is a high level of agreement about their properties. Because of this we can be quite confident that iPS cells are real. But pluripotent cells probably do not exist in the normal organism after the earliest stages of embryonic development.

Chapter 6
Current therapy with tissue-specific stem cells

Haematopoietic stem cell transplantation

Haematopoietic stem cell transplantation is not a term that is immediately recognizable to the general public but it is actually the same as the much more familiar 'bone marrow transplantation'. The rather long phrase haematopoietic stem cell transplantation (HSCT) is now preferred because it covers not just transplantation of bone marrow itself but other types of transplant where the blood-forming (haematopoietic) stem cells of the graft come from non-marrow sources such as peripheral blood or umbilical cord blood. Worldwide, about 50,000 HSCTs are carried out each year making this overwhelmingly the most important type of stem cell therapy in current practice. Most are done for treatment of cancer, mainly lymphomas and leukaemias, with about 5 per cent for treatment of non-malignant blood diseases and a few other conditions.

The history of HSCT goes back to the development of the atomic bomb in World War 2, and the corresponding interest in the effects of radiation on the body. It was quickly realized that radiation killed people and animals because it destroyed dividing cells, and that the haematopoietic tissue of the bone marrow was the most sensitive, followed by the lining of the gut. Secret research work at the time showed that dogs could be rescued from a fatal dose of radiation by shielding part of the bone marrow, or by a bone marrow infusion. After the war, this type of work was continued and the protective effect of bone marrow grafting was ascribed to a substance, perhaps DNA, or some unknown hormone. But suspicion also grew that transfer of live cells might be responsible. In 1955, Joan Main and Richmond Prehn from the National Cancer Institute in Bethesda, Maryland, found that mice which had been lethally irradiated, and then rescued from death with a bone marrow graft, could subsequently accept skin grafts from the same genetic strain of mouse as provided the marrow graft (Figure 14). The work of Peter Medawar and colleagues showing transfer of tolerance by injection of cells into newborn mice had recently been published, so this study made a role for cells seem more likely. The next year, a group led by J. F. Loutit at the Medical Research Council Radiobiology Unit at Harwell, England, carried out a critical experiment in which lethally irradiated mice were rescued with a graft from another strain carrying a visible chromosome abnormality. Once the hosts had recovered, it was found that all of the cells of their blood were of the genetic type characteristic of the donor, so the rescue of mice from the radiation was correlated with, and was probably caused by, the transfer of live cells.

This scientific foundation set the scene for the first attempts at clinical bone marrow transplantation. It was known that high doses of radiation were effective against leukaemia, which is a cancer of blood cells. But it seemed that the dose of radiation that could eradicate the cancer completely would also kill the patient, because of damage to the normal dividing cell populations in the renewal tissues. Among the renewal tissues, the bone marrow is the most sensitive, so the logic was that it should be possible to treat the patient by giving them a lethal dose of radiation to kill the cancer, and subsequently rescue the patient from death by infusion of healthy bone marrow.

Much of the early work was undertaken by E. Donnall Thomas at the Mary Imogene Basset Hospital in Cooperstown, New York,

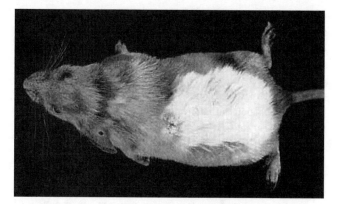

14. A historic skin graft from a white to a tabby mouse. This graft survived because the mouse had previously received irradiation with a lethal dose of X rays and a bone marrow graft from the same inbred strain as the skin donor

an affiliate of Columbia University; and later at the Fred Hutchinson Institute for Cancer Research in Seattle. Thomas was a recipient of the 1990 Nobel Prize for Medicine in recognition of his pioneering work to develop bone marrow transplantation. However, this was no easy task and the early work showed that allogeneic (i.e. between one person and another) bone marrow transplantation was going to be difficult. Until the late 1960s, all of the patients who received an allogeneic graft died of complications associated with the treatment. Although it was not fully recognized at the time, most probably fell victim to a complication called graft-versus-host disease, which is a kind of graft rejection in reverse. Usually the host does not reject a bone marrow transplant because the radiation or chemotherapy given to destroy the cancer has also disabled the host's own immune system. But the graft contains many *T lymphocytes* which will perceive their new surroundings as 'not self' and set about destroying them. During the 1950s and 1960s, a small number of grafts between identical twins showed that the technique could work if the donor was genetically identical to the host. Also, during this early period of experimental therapy, the methods for nursing bone marrow transplant patients were improved and concurrent experiments on dogs clarified what was happening to the patients in terms of graft-versus-host disease. Animal experiments were also critical for unravelling the complexities of the HLA system, which controls immune recognition and which is the principal determinant of the ability of T lymphocytes to recognize other cells as self or not-self.

By 1968, the ability to perform *in vitro* tests on the donor and host lymphocytes to ensure a good HLA match, plus the improved supportive care and antibiotics, began to yield success. The first successful allogeneic transplant was performed by Dr Robert Good in Minnesota in 1968. This involved treatment of a baby with an inherited immune deficiency, using marrow from a healthy sister who was well matched for HLA alleles but did not share the gene variant causing the disease. This patient is still alive today (2012) and still carries the blood and immune system of his donor. Subsequent to this, the immunosuppressive drugs began to appear and for the first time it started to be possible to give effective treatment for graft-versus-host disease.

Modern HSCT differs in various ways from the original. For the treatment of leukaemia, the actual rationale for the procedure has somewhat changed. Instead of the cancer being destroyed by the radiation or chemotherapy and the patient then being rescued by the marrow graft, the main purpose of the graft now is actually to destroy the residual tumour cells. This is because the lymphocytes in the graft mount a particularly effective immune attack on tumour cells which is called the 'graft-versus-leukaemia' effect. The mechanism is poorly understood, and it cannot entirely be separated from the dangerous graft-versus-host effect, so may actually be an aspect of the same thing.

In routine practice, HLA matching is expressed as a score out of 8. There are two main gene groups, ABC and DR, and each is inherited as a copy from the mother and from the father. Each parent carries different versions of ABC and DR (called 'haplotypes' as they are not simple alleles) and passes one haplotype of ABC and one haplotype of DR to any given child. This means that any two siblings will have a one in four chance of having inherited the same haplotypes from their parents and being a perfect match to each other. The chance of a perfect match in the general population is very much lower and as only about 20 per cent of patients have a perfect match sibling, grafts are usually performed with less good matches than 8/8.

Basic science made a further contribution in the 1980s leading to the availability of haematopoietic growth factors. These were discovered as proteins that made the various transit amplifying cells of the haematopoietic system multiply in tissue culture. When the biotech industry got going in the 1980s, haematopoietic growth factors were among their earliest products. One of them, called G-CSF, turned out to be useful for mobilizing haematopoietic stem cells out of the bone marrow and into the peripheral blood. This meant that it became possible to isolate HSCs from the blood of the donor rather than subject them to the rather painful procedure for harvesting bone marrow, which involves inserting a large needle into the iliac crest of the pelvis. However, for an unknown reason, cells prepared this way carry a higher risk of graft-versus-host disease than those isolated direct from the bone marrow, so both methods remain in use.

Blood from the umbilical cords of newly delivered babies is also now used for transplantation since it turns out to contain a much higher content of haematopoietic stem cells per unit volume than adult peripheral blood (Box 6). The severity of graft-versus-host disease following a cord blood transplant is less severe than following an adult transplant, even given the same level of HLA mismatch. Once again, the reason for this is not known. Despite its advantages, the quantity of HSCs found in one umbilical cord is not sufficient to treat an adult patient so they are either used

Box 6: Banking of umbilical cord blood

The utility of umbilical cord blood for haematopoietic stem cell transplantations has led to a new industry of cord blood banking. The blood can be frozen in liquid nitrogen and can be thawed at a later date in viable form, in the same way as tissue culture cells.

Many parents now pay to freeze their own baby's cord blood, 'in case' of needing it for a transplant later in life. While the sentiment behind this is understandable, private banking is much less efficient, as well as more expensive, than donating to a public cord blood bank. A public bank can match samples to patients worldwide so there is a good probability that the blood will actually be used to save a life. By contrast, the vast majority of cord blood banked privately will never be used and will end up being discarded.

to treat children, or more than one reasonably well matched unit needs to be found for each patient.

Because allogeneic HSCT is still extremely hazardous, with at least 10 per cent treatment-associated mortality, it is not prescribed for any but the most lethal conditions. In theory, replacing the entire haematopoietic system would be beneficial for a whole host of autoimmune diseases, for example type 1 diabetes, but the severity of the treatment makes this impractical. One method of reducing the risk of the treatment is to do only a partial marrow ablation in the recipient, in other words reducing the chemotherapy or radiation treatment so that it kills some but not all of the stem cells in the host marrow. It is necessary to kill some, otherwise there are no vacant niches for the graft cells to occupy and the graft will not take, but if the treatment falls short of killing all of the HSCs then it is more likely that the patient will survive, especially if they are over age 60. In terms of cancer treatment, despite the fact that partial marrow ablation will not destroy the tumour entirely, it is felt to offer the better chance of survival, given the favourable graft-versus-tumour effect of the graft.

Just over half of all HSCTs are autologous, meaning that the HSCs are harvested from the patient him-or herself, then the treatment is given, then the graft cells are reinfused. This follows the original rationale of replacing the marrow after lethal irradiation or chemotherapy. It has the considerable advantage that the graft is a perfect match so there is very little chance of graft rejection or graft-versus-host disease. But for treatment of lymphomas and leukaemias it has two disadvantages. Firstly there may well be tumour cells lodged in the bone marrow itself which will miss the treatment and will reintroduce the cancer following the graft. Secondly, because the graft is autologous there is no graft-versus-tumour immunological attack, which seems very effective at removing small numbers of tumour cells that survive the treatment and is now the main rationale behind using HSCT for the treatment of leukaemia. For these reasons, autologous HSCT is mostly used, together with radio and chemotherapy, for the treatment of solid tumours.

HSCT is also sometimes used to treat non-malignant diseases. One category is genetic diseases of the blood or the immune system that are severe enough to cause early death, such as the combined immunodeficiency suffered by the individual treated by Robert Good in 1968. Another is genetic enzyme deficiencies. Particularly if the enzyme is extracellular, or if can be taken up by host cells in functional form, then a haematopoietic graft can be a means of delivering it permanently in an effective dose. Because the disease is genetic in origin, an allogeneic graft has to be carried out in order to deliver HSCs carrying the normal version of the gene, and because of the risk of the procedure it is only contemplated where the disease is itself lethal.

Gene therapy with HSCT

For genetic diseases where the molecular nature of the defect is known it is possible to use an autologous HSC graft as a vehicle for gene therapy. This is done by correcting the genetic defect in the cells while they are out of the body and then reintroducing the 'cured' cells into the patient, following a partial ablation of the marrow. Because the HSCs last a lifetime, and can repopulate the entire blood and immune system, this is a very effective method for introducing genes into the body. A small number of individuals have actually been cured of severe combined immunodeficiencies using this method. Unfortunately, in a few cases the treatment itself turned out to induce leukaemia or lymphoma. The reason for this is that the viruses used to insert the missing genes have a tendency to insert into host DNA next to active genes. Sometimes these may be oncogenes, genes which will make the cells cancerous if they are inappropriately activated. The risk of this complication had been considered rather low on the grounds that there is a lot of genomic DNA to integrate into and the number of oncogenes is quite small. In fact the risk was considered to be about one in a million. However, if genes are introduced into a population of bone marrow-derived cells, even though the number of stem cells is small, there may easily be a million dividing cells of the transit amplifying type, which are capable of being mutated to a cancerous behaviour by the overactivation of an oncogene. In other words, putting genes into such a cell population involves more than one million independent genetic insertion events, each with an oncogenic risk of one in a million. This means that the risk per graft is actually quite large. The publicity given to this problem has impeded the development of otherwise successful gene therapy procedures quite substantially, because it is now necessary to find methods to insert genes into specific locations guaranteed to be non-hazardous, and this is technically very difficult.

Other real stem cell therapy

Although the HSCT is overwhelmingly the most important type of current stem cell therapy, there are a few other examples of real stem cell therapy that have proved successful and are used on a small scale. By 'real' I mean that the cells in question are genuine stem cells and that they actually replace the dead, damaged or non-functional cells of the target organ. There is in addition a large amount of aspirational stem cell therapy where the nature of the grafted cells is uncertain and where they do not become functional cells in the target organ.

Epidermis

The first application is the use of epidermis cultured *in vitro* to treat victims of severe burns. Burns will destroy the skin and leave an open area which is very liable to infection, and, if large, can allow the loss of dangerous volumes of fluid. Burns are frequently treated by autologous skin grafts, where pieces of epidermis are moved from another site on the body to cover the wound. This enables epidermal regeneration to occur from many small pieces of graft placed across the damaged area, such that the gap is closed much more quickly than would occur naturally. The real problem arises when the burns are so extensive that there is not enough healthy skin left to provide grafts. Sometimes allogeneic skin grafts are given, sometimes grafts of animal skin. These measures can provide a temporary protection from infection and fluid loss but they are, by their nature, only temporary because they are rejected after a few weeks.

An ingenious solution was found by Howard Green of Harvard Medical School in the 1970s, who devised a method for the *in vitro* culture of epidermal stem cells. This used feeder cells, rather similar to the later method adopted for growing embryonic stem cells. When provided with the feeder cells and a suitable medium, stem cells from the basal layer of the epidermis will grow in culture and will form a stratified tissue layer quite similar to the epidermis *in vivo*. Growth does not proceed indefinitely but it is sufficient to expand a small biopsy to a large enough area to provide grafts for the whole body in about 3 weeks. Green pioneered the use of this technique to provide grafts for patients with very extensive burns (Figure 15). While it has never become a major activity, a small number of such grafts have been carried out over the years and they have undoubtedly saved several lives.

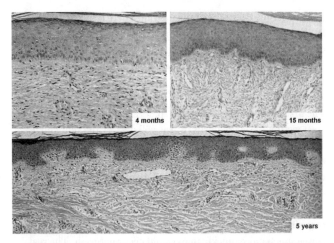

4 months

15 months

5 years

15. **Microscope section showing grafts of skin to a burn victim. This skin was produced by culture of epidermal stem cells from the individual patient. After 4 months the graft has become a well-structured stratified epithelium. After 5 years characteristic dermal ridges are visible. However, hair follicles and sweat glands never develop**

One reason why this therapy has never taken off is, thankfully, that such severe burns are very rare in wealthy countries. Extensive burns are regrettably much more common in developing countries where open fires or paraffin stoves are often used for cooking, and where standards of electrical wiring in dwellings can be rather low. But developing countries do not possess the clinical research centres with the sophistication to undertake stem cell therapy of this sort.

Another reason is that although the grafted epidermis survives indefinitely and gradually integrates well with the underlying dermal layer of the skin, it forever lacks hair follicles and sweat glands. These structures lie in the dermis but they are derived from the epidermis. They each have their own population of stem cells and neither hair follicles nor sweat glands arise spontaneously from the epidermis generated by *in vitro* culture. In fact, the relationship between these structures is the other way round, as the stem cells of the hair follicles can populate the surface epidermis if the adjacent epidermis is destroyed. A lack of hair follicles may be tolerated, indeed many women are constantly removing hair from large body areas, and many men find it possible to survive with a bald head. But life without sweat glands is very unpleasant, as they are critical for regulating body temperature. The absence of sweat glands has been compared to living wrapped in a plastic bag and this must reduce the quality of life for the survivors of this type of stem cell therapy.

An interesting aspect of the epidermal stem cell culture story is that this therapy has been offered by a small number of hospital-based teams who for many years fell below the radar of the regulatory authorities. The cells were cultured under normal laboratory conditions using mouse feeder cells and animal-based products such as the foetal bovine serum commonly used for cell culture. Such methods were quite normal when the activities were started in the 1970s but by the 2000s the regulators started to notice what was going on, and to insist that cells were cultured in special clean rooms dedicated to the purpose, that good manufacturing practice (GMP) should be followed during all steps of the procedure, and that no animal cells or products should be used. The corresponding difficulties and expense have further reduced the level of this small scale but life-saving activity. Interestingly, no patient is known to have suffered any ill effects from the unregulated epidermal stem cell therapy offered in the first 30 years, which suggests that the safety fears relating to animal-derived viruses being introduced during culture are probably exaggerated.

Apart from their use to treat severe burns, epidermal grafts can also be used as vehicles for gene therapy. This application relates to genetic skin diseases where some specific protein is lacking. One example is a form of epidermolysis bullosa in which a subunit of the extracellular protein laminin is missing. This causes a loss of the epidermis because its attachment to the underlying dermis is very feeble, so patients constantly suffer from open wounds which become infected and painful. Despite ferocious regulatory scrutiny of the procedure, involving, as it does, both stem cell culture and gene therapy, one patient has been successfully treated in Italy by the team of Drs Michele de Luca and Graziella Pellegrini. The epidermal stem cells are obtained from a biopsy from a normal skin region and are expanded in culture. The missing gene is then introduced into the cells using a virus. Sheets of multilayered epidermis containing the missing gene are cultured and are grafted in place of the individual's own epidermis. This particular procedure proved successful. Despite that fact that the synthesis of the missing laminin subunit is not regulated in the same way as it would normally be, it seems that the protein can successfully participate in the formation of the complete, three-subunit, laminin protein and alleviate the disease.

Limbus

There is yet another small area of real stem cell therapy. At the front of the eye lies the cornea, a transparent layer allowing light into the pupil. In certain diseases the cornea can becomes opaque, resulting in blindness. This is a type of blindness that can easily be corrected, by means of a corneal graft from a human cadaver donor. Much of the cornea is composed of extracellular protein of low immunogenicity but there is a cell layer on top and underneath. Because systemic immune suppression is not usual for corneal grafts these cell layers may become rejected by the immune system of the host. However, the outer, epithelial, layer is regenerated from the host, because it is normally continuously renewed from a surrounding annular zone called the limbus. So long as the host limbus is intact and functional, a donor cornea will eventually be resurfaced with healthy cells. But a problem arises if the limbus is not intact, for example if the front of the eye has been destroyed by a splash of caustic alkali. In this situation, the grafted cornea will be overgrown not by a benign transparent epithelium, but by opaque tissue from the outer part of the eyeball (the conjunctiva), which will once again cause blindness.

Dr Michele de Luca from the University of Rome, and more recently the University of Modena, in Italy, has spent 30 years developing a method in which cells are harvested from the limbus of the healthy eye, expanded in culture, and grafted to the position of the limbus in the affected eye. If damage to the cornea is slight, this procedure alone can restore a healthy epithelium and effective vision. If the cornea is damaged, then it is replaced with a grafted cornea as well as a limbal graft, and the limbal graft provides the stem cell reservoir for it to be resurfaced by transparent corneal epithelium. It also inhibits resurfacing from the opaque conjunctiva of the host, which would otherwise occur. This method has proved very successful and several dozen patients have had their vision restored. However, at present allografts are not very successful and so the method requires the presence of a viable limbus in the other eye to serve as the source of limbal cells for expansion.

Stem cell therapy of the CNS

A Californian company, Stem Cells Inc., has produced a line of human neuronal stem cells and is using this in trials for therapy of several conditions affecting the central nervous system (CNS). Phase 1 clinical trials have been completed for two rare genetic diseases, Batten disease and Pelizaeus-Merzbacher disease. The first is a disorder in which the absence of a specific protein leads to the accumulation of lipofuchsin (a protein-lipid complex) deposits, particularly in the brain. The result is gradually developing disorders of vision, speech, behaviour, and eventually many other problems. The rationale of the treatment is metabolic rescue by supply of the missing protein, rather than repopulation of a damaged brain with neurons and glia from the graft. Thus, it can be compared with the type of HSCT used to treat enzyme deficiencies. Pelizaeus-Merzbacher disease is a sex-linked genetic disorder, meaning that it occurs in boys whose mothers are carriers. It involves a loss of production of one of the main components of the myelin sheaths of nerve fibres, and leads to a wide spectrum of neurological diseases ranging from mild to very severe. The rationale here is the formation of remyelinating cells from the graft that may be able to repair the damage. In 2011, a phase 1/2 trial of the same cells was also initiated for treatment of chronic spinal injury. Although the CNS has a degree of protection from the immune system, immunosuppression is still necessary for this type of therapy.

Stem cell therapy of the heart

At present there is no real stem cell therapy of the heart. That is to say, no one has succeeded in grafting cells into the heart that will repopulate it and replace damaged cardiac muscle. However, over about 10 years there has been a remarkable level of activity in many centres in which cells of various sorts are grafted into diseased hearts. This activity lies somewhere in between real and aspirational stem cell therapy. The good aspect is that many patients have been enrolled into well-controlled clinical trials, so there is a good understanding of the efficacy of the treatments. The bad side is that the rationale for the treatments is rather weak and it is hard to tell what is happening.

The rationale derives from a burst of laboratory experiments on animals around the turn of the 21st century indicating that grafts of bone marrow into irradiated hosts could repopulate many organs. We expect a bone marrow graft to repopulate the blood and immune system, but these studies reported repopulation also of organs such as the brain, the liver, the pancreas, and the heart. This phenomenon became mis-named 'transdifferentiation'. Careful investigation eventually indicated that virtually all of the apparently positive results were due to various artefacts and that transdifferentiation does not occur to an appreciable extent. This type of transdifferentiation may actually occur to a very limited degree. For example there have been post-mortem studies of bone marrow recipients in which the donor was male and the host female, and it is possible to identify male cells in various tissues because of the presence of a Y chromosome in the cell nuclei.

However it certainly does not occur at a high enough frequency to be of any clinical utility.

Unfortunately the details of this rather arcane dispute and the careful animal experiments that were required to settle it were not fully absorbed by clinicians, some of whom eagerly started conducting grafts of autologous bone marrow into the hearts of patients who had suffered heart attacks, or had developed heart failure. The rationale was that 'bone marrow contains stem cells that can repopulate any tissue in the body'. To be fair, the rationale was at least as strong as that which had prompted the early trials of bone marrow transplantation, and the procedure was much less lethal, so could be undertaken under modern regulatory conditions in which Institutional Research Boards scrutinize every proposal very thoroughly in advance to ensure that the balance of risk and benefit is reasonable; that the experiments are well designed; and that all patients have given informed consent.

The results of a large number of trials can be summarized by saying that cardiac function does improve, but only slightly and usually not for very long. For example, a 2 per cent increase of cardiac output lasting a few months would be typical. It is not possible to make very detailed post-graft studies in human patients, but corresponding animal experiments indicate that all the cells die fairly soon after grafting. So the benefits, if they exist, are not due to repopulation of damaged heart muscle. They might perhaps be due to stimulation of new blood vessel formation, or to a reduction of inflammation, or to some other ill-defined immune modulation. Such mechanisms are generally called 'paracrine effects', a vague term indicating effects due to some unknown substances released from the cells.

A further difficulty with this type of procedure is that there is a substantial placebo effect in the controls. The cases where similar injections are made into the diseased heart, but with no cells, show almost as much benefit as the cases where cells are actually injected. So the non-specific mechanisms alluded to above may be real, but may not require actual cells at all. This example of 'stem cell therapy' of the heart is perhaps the one in which the division of views between clinicians and scientists is at its widest. Scientists generally regard the whole enterprise as pointless, lacking rationale and producing no valid results. Clinicians cite the statistically significant positive results and justify the activity on the grounds of benefits to patients, while admitting that the mechanism is not understood.

Chapter 7
Realistic and unrealistic expectations

When we consider our lot in terms of medicine, we are usually very thankful. We look back in horror and contemplate the huge infant mortality and all too frequent maternal mortality in times before the twentieth century. Fortunately our present level of technology has reduced these risks to a level where they do not loom large for most people. Nonetheless, in some respects we are still in the dark ages. Future generations will surely look back on us, and marvel that we could have lived so acceptingly with permanent paralysis as a result of spinal injury, or permanent loss of limbs resulting from severe tissue damage, or inevitable death from heart failure or from many forms of cancer. All those involved with the biomedical sciences are confident that one day we shall be able to regenerate missing structures and also that there will be cures for heart failure, diabetes, cancer, and neurodegeneration, but we have little real idea when or in what way these things will be achieved. Success probably will come in the end but it is likely to come more slowly that most people expect. Inflated expectations are highly prevalent because we live in a world where money can be acquired by attracting attention and by making promises, and any promise of new cures for serious and debilitating diseases is bound to attract a lot of attention. The hype has been further amplified by the ethical debate over human ES cells, which led the proponents of stem cell research to promise very rapid development of very radical cures. It is further increased by the belief of many politicians in many countries that stem cell therapy is the 'next big thing' after computers, and its development can, in some way, rescue their uncompetitive and failing economies.

Hopes and realities

Nowhere has the hype been greater than in California where the California Institute for Regenerative Medicine (CIRM) was set up using $3 billion of public money, raised from state bonds. The motivation was an attempt to circumvent the federal funding restrictions on human ES cell research introduced by President Bush in 2001 (see Box 7), by providing state funding instead. California is always pleased to do things that annoy Washington, and its scientists were ready to use all of their persuasive power to help the endeavour get underway. As a result of all the blandishments, the Californian public really believes that new cures will come in just a few years, and the politicians really believe they will get their money back in taxes from new profitable companies built on the technology. The scientists tend to be somewhat less optimistic in private than they are in public. Some of those who work with human pluripotent stem cells do not think that there will ever be cell therapies based on their use. Instead, they argue that the value of human pluripotent stem cells is the possibility of studying human embryonic development without embryos, or of screening drugs for safety using hard-to-obtain human cell types such as cardiomyocytes. However, the Californian public did not subscribe $3 billion for such activities, they wanted cures, and they expected cures in a few years. Because of this it is widely feared that there will be a degree of backlash when the cures turn out to be slower coming and more limited in scope than had been hoped.

Despite this fear, a backlash may not be inevitable. The technologies being developed are very powerful and there will be real progress achieved as a result of the CIRM initiative.

Box 7: Restrictions on stem cell research in the USA

Contrary to popular belief there is no federal prohibition of any type of stem cell research in the USA. What does exist is a restriction on the way in which public money may be spent, specifically research grants from the National Institutes of Health (NIH).

The main limitation is a 'rider' attached by Congress every year since 1996 to its main appropriation bill for Health and Human Services, which prohibits spending of public money on any activity involving the 'destruction' of human embryos, which is taken to include the generation of ES cell lines from human embryos.

In addition to this, President George W. Bush introduced an Executive Order in 2001 which limited public funding for research using human ES cell lines. Research on a limited number of existing lines was permitted, but new lines were excluded. In 2009 this Order was replaced by one from President Obama which allowed public funding for work on new ES cell lines so long as they were established following stringent ethical guidelines.

However, it is still not possible to make new human ES cell lines using public funds. Either they must be made using private funds or obtained from abroad. This position is unlikely to change in the foreseeable future.

Opinions differ on how damaging the restrictions have been to the US research effort. NIH funds do support the bulk of biomedical research, and private funds (companies, charities, individual philanthropy) are relatively limited, so the public funding restrictions are a significant problem. On the other hand, the countries with more liberal regimes (UK, Singapore, Australia, Sweden) have not really forged ahead of the USA during this period. Perhaps the main damage has been to the perception of the USA's position as the world leader in biomedical research rather than to its actual performance.

The problem is that the whole of the biomedical sciences are producing less in terms of real commercial or clinical outputs compared with what they produced in the 1940s and 1950s, when the level of investment was much smaller than it is today. The reasons for this are hard to dissect, but they must include the fact that most of the easy stuff has been done, and what remains is very difficult, with ferocious scientific and technical problems to be overcome. In addition, there are three other problems that slow down the introduction of new therapies.

Firstly, the process of gaining approval to introduce any new therapy has become more and more cumbersome, and takes a very long time. Secondly, some of the diseases in question have other, existing therapies, which can be quite successful. Any new therapy has to be significantly better than the current best in order to be taken up. Finally, there has been a lack of venture capital investment because the financial returns are considered insufficiently certain to materialize. In particular, the promise of personalized cell therapy inherent in stem cells, which motivates many scientists, is considered by most analysts to be prohibitively expensive and so it is unlikely that this technology will see major private investment in the near future.

Because the FDA and other regulatory agencies are often criticized for slowing down the introduction of new therapies, it is worth remembering that they exist in response to public demand and as a response to past disasters. In this context one example of aspirational stem cell therapy, normally considered as 'safe', is instructive. It was described in Israel in 2009. A young boy suffering from ataxia telangiectasia (a rare genetic disease involving a defect of DNA repair, leading to severe motor problems) had been taken to Moscow on three occasions for stem cell therapy which consisted of injections into the brain and spinal cord of neurogenic cells derived from human foetuses. A few years later he was found to have multiple tumours, containing both neurons and glial cells, which originated from the donor cells.

Normally such aspirational therapy is relatively safe, if ineffective. This is because an allogeneic graft will be rapidly rejected in the absence of immunosuppression and will probably not do too much harm. But as well as the neuronal problems that it causes, ataxia telangiectasia involves a degree of immune deficiency, and this, together with the relative protection of the central nervous system from the immune system, probably explains why the grafts from the human foetuses were able to survive and to seed the tumours.

Lessons from HSCT

Since haematopoietic stem cell transplantation (HSCT) is so far the one major success story of stem cell therapy in clinical practice, it is worth carefully considering the lessons from its development. Some of them are surprising and may even be unwelcome to various professional groups.

Firstly, it is remarkable how little scientific understanding there was of the haematopoietic system when the first clinical forays were made in the 1950s. It would be decades before isolation of haematopoietic stem cells, or characterization of the stem cell niche, or understanding of the cell lineage, or isolation of the various haematopoietic growth factors. It is often argued today by scientists that it is 'far too early' to contemplate clinical trials of cells made from pluripotent stem cells, or neurospheres, or tissue engineered products. Given the huge mortality from the early bone marrow transplants it probably really was too early in the 1950s, but even by the time success started to be achieved in the late 1960s, the scientific understanding of the haematopoietic system was still rudimentary. It was not until 1988 that the isolation of mouse HSCs was achieved (by Irving Weissman, of Stanford University), with human HSCs following a few years later. What led to success was not detailed understanding of the system but advances in other areas, principally the understanding of the HLA genetic system controlling rejection, the development of practical methods for tissue matching, and the discovery and practical application of immunosuppressive drugs.

Secondly, the technology of haematopoietic transplantation was developed mostly in the absence of external regulation, which only came into force in the 1980s for new clinical procedures. The enormous mortality of the early years was perhaps justifiable since the patients all had fatal diseases and so a significant risk could reasonably be taken in the search for a cure. But everyone knows that this type of approach could not possibly be followed today. Indeed, the view is often voiced that the FDA, and its counterparts in other countries, would simply not allow the necessary risks to be taken and so the technology could not have been developed at all.

Thirdly, it is notable that the basic rationale of the procedure has changed. Originally the rationale was that eradication of leukaemia needed a lethal level of radiation and the patient would be enabled to survive the lethal dose by replacing the bone marrow. Now the rationale is mostly that an allogeneic graft has a powerful anti-tumour effect, probably a by-product of the graft-versus-host effect, and this is its chief therapeutic value. Accordingly, it is increasingly common for the treatment to be designed to produce less than 100 per cent ablation of the host bone marrow.

Furthermore, HSCT has become extended into other areas which were not originally foreseen. For example, recently Drs John Wagner and Jakub Tolar at the University of Minnesota have used such grafts to treat a genetic skin disease: a type of epidermolysis bullosa in which the extracellular protein collagen 7 is missing. This is more severe than the form of the disease mentioned in Chapter 6. The lack of collagen 7 means that the epidermis of the skin is poorly attached to the underlying dermis and falls off very easily. Babies born with the condition suffer massive injury, severe infection and constant pain, and do not survive long. It turns out that HSCT can improve the condition markedly, probably because some of the numerous cell types produced by the haematopoietic system themselves secrete collagen 7, and enough of it finds its way to the junction of epidermis and dermis to rescue the defect. This is an important innovation to treating a very severe disease, but the simple mechanism was not expected or predicted before the event.

Fourthly, despite huge advances in cure rates for leukaemias and lymphomas, the allogeneic haematopoietic graft remains an extremely aggressive procedure with a high treatment-associated mortality. This means, notwithstanding the above example, that it is not suitable for treating a wide range of other diseases in which the cause lies in some defect of the blood or immune system, because the risks could not be justified for conditions that are not rapidly fatal.

Finally, it is worth considering the economic impact of HSCT. Politicians like to think of stem cell research as driving economic growth. Their usual vision of technology transfer is that new discoveries are made in laboratories and are then put into a pipeline leading to inevitable commercial exploitation. At some stage they will be patented to generate valuable intellectual property. Then a product will be developed under patent protection, a company will be set up, sell the product for a high price and pay a lot of taxes to the government. But this attractive vision does not correspond well to reality. Most of the discoveries of academic scientists have no commercial value. For those that appear to have some value, the vast majority are discarded during the development process for all sorts of reasons. When we think about HSCT, the first consideration is that it is not a product, it is a service. HSCT has certainly generated some wealth, but in rather indirect ways hard to predict in advance. In countries such as the USA with largely private health provision, the service can be charged for, and it is very expensive. HSCT has also generated many ancillary products and services such as machines for fractionating cells from marrow or blood, antibodies for identifying cells, and the immunosuppressive drugs and antibiotics used to treat the patients. The production and marketing of all of these things has produced wealth, but they are all ancillary to the main procedure. In countries with a state-funded medical system, such as the UK, introduction of a new (free) service is likely to be seen as a cost, not as a benefit, from an economic standpoint, so it could even be perceived as having a negative economic impact.

The overall impression is that in all the above respects, the story of HSCT is one that could not have been foreseen in advance by any analyst, however gifted. Even when it is examined in retrospect, with the benefit of hindsight, it fails to fit most current visions of stem cell research. Scientists usually believe that a hugely detailed understanding of the mechanism of the natural process is necessary before embarking on any therapy based upon it. The story of HSCT clearly shows that this is incorrect. Regulatory authorities do not accept that their activities impede the development of new medical technology. But it is very doubtful whether current regulators would ever sanction lethal whole body radiation or equally lethal allogeneic grafts as new therapies today. In terms of economic benefits, politicians would not see new services provided by state-funded hospitals as generating wealth for their countries, they would see them as irksome new costs.

The more positive lessons to be drawn are perhaps as follows. Firstly, there is quite a long time delay between the minimal scientific understanding on which a new therapy might be based, and actual practical success. In the case of HSCT it was about 20 years between the discovery that bone marrow cells cured radiation sickness in mice, and the establishment of bone marrow transplantation as a treatment for humans with any hope of success. Nowadays, the much more stringent regulatory oversight means that the time delays will be longer rather than shorter. This supports the contention of many scientists that it is still 'early days' for stem cell therapy. Secondly, although not much scientific understanding is required to invent a new treatment, you certainly do need scientific knowledge to be able to measure the outcomes accurately. So a whole raft of basic science was actually needed to make HSCT possible. This included the understanding of the types of cell in the blood, and how to identify and count them; the nature of leukaemia and lymphoma as uncontrolled proliferation of haematopoietic cells; their origin from precursors in the bone marrow; the nature and measurement of graft rejection, essential for discovery and development of immunosuppressive drugs; and the genetic basis of graft rejection, essential for tissue typing. Although you don't need to know a lot about how the haematopoietic system works to invent bone marrow transplantation, you do need to know a lot of other stuff to be able to measure what is happening, so that you can assess the course of the disease and the response to therapy. In other words, basic science is necessary after all, but not in the way usually expected. In fact you can never really predict which parts of it will be necessary for which applications.

The future

In the long run the sky is the limit, and most biomedical scientists feel that the huge advances in understanding and ability to manipulate cells and genes must eventually lead to some big innovations. But in the coming 10 years progress with stem cell-based therapies is likely to be more modest.

The first clinical trials of pluripotent cell-derived cells have already started, for spinal injury and macular degeneration. Within a few years there will be some new clinical trials involving grafts of pluripotent cell-derived beta cells for diabetes, cardiomyocytes for heart disease, and perhaps dopaminergic neurons for Parkinson's disease. For tissue-specific stem cells, there will continue to be incremental improvements of current HSCT for cancer and extensions of its uses in the area of genetic deficiency diseases. The less lethal protocols may become more widely used for some autoimmune diseases. Further trials will undoubtedly be carried out using neuronal stem cells of various sorts, and the existing therapies based on transplantation of epidermal and limbal stem cells will be widened to include using these cells as vehicles for delivery of gene therapy.

In terms of advances in science that are likely to prove transformative, the possibility of expanding HSCs *in vitro* is being actively pursued and this will be very useful if it can be used to generate grafts with a higher proportion of HSCs and fewer other cell types. There is also the prospect of generating HSCs, or blood products such as platelets, from pluripotent stem cells, which could ease the donor problems for HLA-matched grafts. Otherwise, probably the main area of interest lies in what is now called 'direct reprogramming'. This means using methods similar to those currently used to make iPS cells to achieve a direct, single step, transformation from the cell of origin, which might be a fibroblast or a white blood cell, to the desired cell type, which might be a neuron, cardiomyocytes or beta cell. As with iPS cell production, specific transcription factors are introduced into the target cells which are then subjected to a suitable selection procedure to obtain the desired cell type. The choice of transcription factors depends on knowledge of the normal developmental pathway to the cell type in question during embryogenesis. Preliminary studies have shown that it is possible to produce neurons, cardiomyocytes and beta cells by this type of method. Interestingly they really do seem to arise in a direct, single step event, without passing through the normal intermediate states found in embryonic development. At present the transcription factors are introduced as genes using integrating viruses, and it will be necessary to find alternative methods not involving gene integration if such cells are to be used for therapy. But this is likely to be achieved fairly soon because many groups have been tackling the same issue in relation to iPS cells. Cells made by direct reprogramming will be a perfect genetic match to the person from which the original cells were derived. Even better, because they have not passed through a stage of pluripotency, there should be no risk of creating teratomas from persistent pluripotent cells in the graft.

The other area of likely technical progress concerns methods of cell reimplantation. Because simple injection seems usually to lead to large scale death of the cells, it is likely that future treatments will involve more sophisticated types of implant made by the methods of tissue engineering. Tissue engineers have developed many scaffold materials for keeping cells in viable condition in three-dimensional arrays. They have also adapted the methods used to make electronic microchips to build up complex three-dimensional structures layer by layer. Future implants are likely to consist of several layers of different cell types, provided with a vascular system which can be connected to that of the host. It will therefore be a graft of a piece of organized tissue, resembling the normal tissue, rather than an injection of a single type of cell. The cells incorporated into these grafts will be differentiated cells made from pluripotent stem cells, or made by direct reprogramming, and if the cost of personalized cell culture can be brought down, they will be a perfect genetic match to the patient.

In the long term the new technology represented by stem cell biology must surely have enormous potential. In fact, it should be seen in a broader context as a technology of regenerative medicine, which embraces gene therapy and tissue engineering as well as stem cell biology. But, as in other areas of life, accurate prediction of the future is quite extraordinarily difficult. So I shall make no more specific predictions, but simply hope that those who have read this book will be able to use their knowledge and judgement to understand the context of news reports about new discoveries, be sceptical about the claims of private stem cell clinics to provide miracle cures, and, if they are involved in business or politics, to contribute to sensible decisions about the funding and regulation of stem cell research in the next few decades.

Glossary

'Adult stem cell': Any type of human **stem cell** that is not an **embryonic stem cell**.

Allele: A specific variant of a **gene**. Each gene is composed of a length of **DNA** and occurs at a specific position (or locus) relative to other genes and to non-coding DNA. The DNA sequence of a gene may differ slightly between different individuals, or between the maternal and paternal chromosomes in one individual. Each of these specific variants is called an allele.

Allogeneic graft/allograft: Transplantation of cells, tissue or organ from one individual to another.

'Aspirational stem cell therapy': A term used only in this book to indicate forms of stem cell therapy with no clear rationale or evidence of effectiveness.

Autologous graft/autograft: Transplantation of cells, tissue or organ to the same individual from which it was derived.

Astrocyte: Type of **glial cell** that forms scars following damage to the central nervous system.

Beta cells: Cells that produce the hormone **insulin** in response to an increase in the concentration of glucose. They are located in the **islets of Langerhans** in the pancreas, along with other **endocrine** cell types.

Blastocyst: The stage of mammalian development when the embryo consists of a **trophectoderm** layer surrounding a fluid-filled cavity and an **inner cell mass**.

BrdU (Bromodeoxyuridine): A **DNA** precursor, similar to thymidine (the T of A, C, T, and G), which is incorporated into DNA when it is synthesized by cells. DNA containing BrdU can be visualized on tissue sections by staining with a specific antibody, showing which cells were making DNA when the BrdU was administered.

Cardiomyocyte: A heart muscle cell.

Cell culture: See **tissue culture**.

Chimaera: An animal composed of two genetically different populations of cells. For example, a mouse formed from an embryo injected with **embryonic stem (ES) cells** consists partly of descendants of the ES cells and partly of descendants of the original embryo cells.

Cloning: Making many genetically identical copies of a molecule (usually **DNA**), or a cell, or a whole organism.

Connective tissue: The tissue in between other structures, such as the dermis of the skin, and various capsules and sheaths, which are largely composed of fibroblasts. The term often also includes bone, cartilage, tendons, and ligaments.

Cytokines: Extracellular regulatory proteins controlling growth, death, differentiation or other cell functions. Cytokines are similar to **growth factors**, but generally are products of cells of the immune system.

Cytoplasm: The part of the cell which is not the **nucleus**. Although it appears as a clear jelly in the light microscope, the cytoplasm consists of a complex set of organized structures which perform most of the biochemical functions of the cell, including **protein** synthesis, secretion, absorption, chemical transformation of small molecules, and so on.

Differentiated/undifferentiated: Relating to cells, a differentiated cell has specific features, usually visible in the light microscope, and a specific function in the body.

DNA (deoxyribonucleic acid): The material that makes up **genes**. DNA is a polymer composed of four types of chemical units, called nucleotides, and represented by the symbols A, C, T, and G. A gene consists of a certain length of DNA that encodes one **protein**. The specific sequence of nucleotides in the gene encodes the sequence of amino acids in the protein that is synthesized when that gene is active.

Ectoderm: The outer of the three **germ layers** typically making up the body of an **embryo** following the **primitive streak stage**. The ectoderm later differentiates to form the epidermis and the central nervous system.

Egg: In this book the colloquial usage is followed, so 'unfertilized egg' refers to a secondary **oocyte**, and 'fertilized egg' to a fertilized oocyte.

Embryo: The early developing stage of a human or animal. Mammalian early embryos contribute parts to the placenta as well as becoming the actual foetus.

Embryoid body: A structure slightly resembling an embryo, but very disorganized in structure, formed *in vitro* by differentiation of an aggregate of **embryonic stem cells**.

Embryonic stem (ES) cells: Cells arising from *in vitro* culture of the **inner cell mass** of a mammalian **blastocyst** stage embryo. In appropriate media, ES cells can grow without limit, or can be caused to differentiate into all the cell types found in the normal body.

Endocrine cells: Cells that secrete hormones into the blood stream, such as the **beta cells** located in the pancreatic islets, which secrete **insulin**.

Endoderm: The innermost of the three **germ layers** typically making up the body of an embryo following the **primitive streak** stage. The endoderm later forms the epithelial lining of the gut and respiratory system.

Epithelial cell: A generic type of cell that normally lines surfaces or makes up glandular tissues, for example keratinocytes (skin cells) or **hepatocytes** (liver cells). In the body they lie on a basement membrane and are attached to each other by specialized junctions. Epithelial cells usually grow as flat sheets in tissue culture.

Feeder cells: Cells in tissue culture that have been prevented from dividing by X-irradiation or by cytotoxic drug treatment, but which remain biochemically active. Feeder cells are able to support the growth of **embryonic stem cells** because of the **growth factors** which they secrete.

Fibroblast: A generic cell type that normally fills spaces and secretes the protein collagen, and has an elongated or stellate appearance. Most **tissue culture** cells are thought to be derived from fibroblasts.

Foetus: A human **embryo** officially becomes a foetus after the first two months of gestation. By this stage, all organs are formed and are maturing, and the visible appearance is dominated by the well-formed limbs.

Gene: A length of **DNA** that encodes a specific **protein**. When a gene is active, it is said to be being **expressed**.

Gene expression: The production of **protein** encoded by a particular **gene**.

Genome: The full set of **genes** for one species of animal or plant. With a few exceptions, each cell **nucleus** contains the entire genome.

Germ cells: Reproductive cells: sperm or **eggs**, or their **progenitors**.

Germ layers: The **ectoderm, mesoderm**, and **endoderm** of the early **embryo** are referred to as germ layers. Not to be confused with **germ cells**.

Glial cells: Non-neuronal cells found in the central nervous system, especially **astrocytes** and **oligodendrocytes**.

Growth factors: Extracellular regulatory **proteins** controlling growth, death, differentiation, and other cell functions. Similar to **cytokines**, but generally are products of non-immune cells. Some growth factors are **inducing factors** active in embryonic development.

Hepatocyte: The main functional cell type of the liver.

Haematopoietic: Relating to the formation of the blood and immune system.

Haematopoietic stem cells (HSCs): Stem cells resident in the bone marrow that generate all the cell types of the blood and immune system.

Haematopoietic stem cell transplantation (HSCT): Therapeutic transplantation of HSCs derived from bone marrow, peripheral blood, or umbilical cord blood.

HLA (Human Leukocyte Antigen): A complex of **genes** encoding the **proteins** responsible for graft rejection. There are several genes in the HLA complex and many possible **alleles** of each. Allogeneic grafts between individuals who are well matched for their HLA types will be tolerated better than poorly matched grafts, and require less immunosuppression for survival.

Immuno-: Relating to the immune system, and in the context of this book especially the parts of the immune system responsible for graft rejection. Thus, *immuno*suppressive drugs can inhibit graft rejection; *immuno*deficient mice can accept grafts of human cells; and *immuno*compatible identical twins can exchange grafts with one another.

Induced pluripotent stem cells (iPS cells): Cells closely resembling **embryonic stem (ES) cells** that are made from normal **somatic** cells by introduction of a small number of **genes** that encode **pluripotent** behaviour.

Inducing factors: Extracellular **proteins** that control embryonic development by causing cells to select a specific pathway of development in response to concentration. Chemically, inducing factors are a subset of the **growth factors** and **cytokines**, and they can have other functions in life after birth.

Inner cell mass: The clump of cells within the **blastocyst** stage of a mammalian **embryo**. It forms the embryo proper as well as some layers of the placenta.

Insulin: A hormone, secreted by **beta cells** of the pancreas, which is essential for allowing entry of glucose into muscle and adipose (fat) cells, and for other metabolic functions.

iPS cells: See **Induced puripotent stem cells**.

Islets of Langerhans: Small structures in the pancreas containing the **beta cells** and other **endocrine** cell types.

Isotope: A single chemical element usually has more than one isotope, each characterized by a different atomic mass. Many isotopes are stable but others are radioactive and decay into other elements with a characteristic half life.

Lymphocytes: Cells of the immune system. They are roughly spherical with a large nucleus and little cytoplasm. They arise from the **haematopoietic stem cells** (**HSCs**) of the bone marrow and are found in the blood. The two major classes are B lymphocytes, which produce antibodies, and T lymphocytes, which are responsible for recognition and destruction of infectious organisms, and also for graft rejection.

Mesoderm: The middle of the three **germ layers** formed from the **primitive streak** stage of the early **embryo**. It will later form the muscles, skeleton, other **connective tissues**, kidney, and gonads.

Mesenchymal stem cells (**MSC**): Cells found in many tissues which can be cultured *in vitro* and which show an ability to differentiate into bone, cartilage, adipose (fat) tissue or smooth muscle.

Mitosis: The process of cell division, comprising segregation of chromosome sets to the daughter cells and the physical division of the cell into two.

Mitotic figures: The arrangement of chromosomes in cells undergoing **mitosis**.

Muscle satellite cells: Small cells lying beneath the basement membrane of muscle fibres, which are capable of self-renewal and generation of muscle progenitor cells, normally following tissue damage.

Neuron: A nerve cell. Neurons can stimulate other neurons, or muscle cells or gland cells, by means of electrical impulses that travel down nerve fibres and cause release of neurotransmitter substances adjacent to the target cells.

Neural stem cells: Cells that can self-renew and also generate the main cell types of the central nervous system (**neurons, astrocytes**, and **oligodendrocytes**). In adult mammals they are found in two parts of the brain: the lining of the lateral ventricles and in the dentate gyrus of the hippocampus. Similar cells are present in **neurospheres.**

Neurospheres: Small clumps of cells that can be grown in suspension culture in a medium containing the **growth factors** EGF and FGF. Neurospheres contain **neural stem cells** and **transit amplifying cells**, and can differentiate into **neurons, astrocytes** and **oligodendrocytes**.

Niche: A microenvironment favourable for the continued survival, proliferation, and function of **tissue-specific stem cells**.

Nucleus: The nucleus of a cell is a roughly spherical structure bounded by a membrane and contains the **genes**.

Oligodendrocyte: A type of **glial cell** in the central nervous system responsible for the production of insulating myelin sheaths for nerve fibres.

Oocyte: A female **germ cell**. A primary oocyte is formed following the last division of its **germ cell** precursors, which in mammals occurs during foetal life. Primary oocytes persist in the ovary throughout the years of fertility. Following hormone stimulation they are released and mature. Fertilization takes place at the stage of maturation known as the secondary oocyte, colloquially known as an 'egg'.

Oncogene: A **gene** that imparts tumour-like behaviour on cells when it is active at an excessive level.

Oval cells: Cells located in the liver that behave as multipotent **progenitors** for **hepatocytes** and biliary cells, two types of liver cells.

Passage: The number of times that a cell line has been subcultured since it was isolated from the organism.

Pluripotent, pluripotency: The ability to differentiate, usually via a sequence of developmental stages, into all of the cell types normally found in the body. 'Totipotency' used to be the term for this but has fallen out of favour.

Post-mitotic: Referring to a cell that has undergone its last cell division (**mitosis**) and will not divide again.

Postnatal: After birth, referring to a juvenile or adult mammal.

Protein: Proteins make up most of the content of a cell and carry out most of its specific functions. They are large molecules, which are polymers of small units called amino acids. Each protein is encoded by a **gene**, and the sequence of amino acids in the protein molecule depends on the sequence of A, C, T, and G in the DNA.

Preimplantation embryo: A mammalian embryo in its early free-living stage, before it implants in the uterus.

Primitive streak: A mass of cells in the early mammalian **embryo** within which cells are moving to form the three **germ layers**, and where **inducing factors** are active to generate the first major subdivisions of the body plan.

Progenitor cells: Cells that are dividing and are committed to form a specific cell type. Unlike **stem cells**, they do not self-renew long term. **Transit amplifying cells** and cells of the early embryo are both examples of progenitor cell populations.

Renewal tissue: A tissue in which the differentiated cells are continuously replaced by cell division from a population of **tissue-specific stem cells**. Examples are the epidermis, the blood, the lining of the intestine, and the testis.

Reproductive cloning: Making an identical genetic copy of an animal, or potentially a human being. In animals this has been done by **somatic cell nuclear transfer** (**SCNT**) into an **oocyte**, or by injection of **iPS cells** into an early embryo, followed by breeding from the resulting **chimaeric** embryo.

Somatic cells: Any cell of the body apart from the reproductive **germ cells**.

Somatic cell nuclear transfer (SCNT): Replacement of the **nucleus** of a secondary **oocyte** by that of a **somatic cell**. If successful, the resulting embryo will have the genetic constitution of the donor nucleus. This is a method for **reproductive** or **therapeutic cloning** of animals.

Stem cell: A cell that persists for the lifetime of the organism and continues both to reproduce itself and to generate **differentiated** progeny.

T lymphocytes: See **lymphocytes**.

Teratoma: Tumour containing a wide mixture of tissue types, often characteristic of those formed from all three embryonic **germ layers**. Teratomas may be spontaneous, usually arising from **germ cells**. In the context of stem cell research, the formation of a teratoma containing a wide range of tissue types is the best criterion for **pluripotency** of **human embryonic stem** (**ES**) **cells** or **induced pluripotent stem cells** (**iPSC**).

Therapeutic cloning: The process of creating an embryo by **somatic cell nuclear transfer** to an **oocyte**, and then using it to establish an **embryonic stem (ES) cell** line.

Tissue culture (=**cell culture**): Growth of cells *in vitro*.

Tissue-specific stem cells: The **stem cells** found in juvenile and adult bodies which are responsible for maintenance of **renewal tissues**.

Totipotency: See **pluripotency**.

Transcription factor: A type of **protein** that regulates the **expression** of **genes** encoding other proteins.

Transit amplifying cells: Dividing cells that are derived from **tissue-specific stem cells** and are committed to **differentiate** after a certain number of divisions.

Trophectoderm: The outer layer of a mammalian **blastocyst** stage **embryo**. It is the first structure to differentiate from the early embryo and becomes part of the later placenta.

Further reading

The public section of the website of the International Society for Stem Cell Research contains some very accessible and informative sections: http://www.isscr.org/public/index.htm

The USA National Institutes of Health website also contains plenty of useful information: http://stemcells.nih.gov/info/

There are many recent books about stem cells. Here are a few:

L. S. B. Goldstein, and M. Schneider, *Stem Cells for Dummies* (Hoboken NJ: Wiley Inc., 2010).
(Introductory but not really for dummies)

C. B. Cohen, *Renewing the Stuff of Life: Stem Cells, Ethics and Public Policy* (Oxford: Oxford University Press, 2007).
(Legal and moral issues surrounding human ESC)

Christopher Thomas Scott, *Stem Cell Now: A Brief Introduction to the Coming Medical Revolution* (New York: Plume, 2006).
(Mostly dealing with law and ethics)

C. Mummery, I. Wilmut, A. Van de Stolpe, B. Roelen. *Stem Cells: Scientific Facts and Fiction* (London: Academic Press, 2011)
(Nicely illustrated science review)

H. Green, *Therapy with Cultured Cells* (Singapore: Pan Stanford Publishing Pte, 2010).
(A rather terse account of work with cultured skin cells)

The following review articles in scientific journals are tougher reading but provide an entrance to the primary scientific literature.

Basic biology

M. Baker (2009) Fast and Furious (about iPS cells). *Nature* 458, 962–5.

C. Crosnier, D. Stamataki, and J. Lewis (2006) Organizing cell renewal in the intestine: stem cells, signals and combinatorial control. *Nature Reviews Genetics* 7, 349–59.

E. Fuchs (2007) Scratching the surface of skin development. *Nature* 445, 834–42.

G. Keller (2005) Embryonic stem cell differentiation: emergence of a new era in biology and medicine. *Genes & Development* 19, 1129–55.

M. Kondo, A. J. Wagers, M. G. Manz, S. S. Prohaska, D. C. Scherer, G. F. Beilhack, J. A. Shizuru, and I. L. Weissman (2003) Biology of Hematopoietic Stem Cells and Progenitors: Implications for Clinical Application. *Annual Review of Immunology* 21, 759–806.

M. I. Koster and D. R. Roop (2007) Mechanisms Regulating Epithelial Stratification. *Annual Review of Cell and Developmental Biology* 23, 93–113.

J. S. Robert (2004) Model systems in stem cell biology. *BioEssays* 26, 1005–12.

H. Shenghui, D. Nakada, and S. J. Morrison (2009) Mechanisms of Stem Cell Self-Renewal. *Annual Review of Cell and Developmental Biology* 25, 377–406.

G. Vogel (2010) Diseases in a Dish Take Off. *Science* 330, 1172–3.

S. Yamanaka (2009) A Fresh Look at iPS Cells. *Cell* 137, 13–17.

X. Yang, S. L. Smith, X. C. Tian, H. A. Lewin, J-P. Renard, and T. Wakayama (2007) Nuclear reprogramming of cloned embryos and its implications for therapeutic cloning. *Nature Genetics* 39, 295–302.

Stem cell therapy

A. Abdel-Latif, R. Bolli, I. M. Tleyjeh, V. M. Montori, E. C. Perin, C. A. Hornung, E. K. Zuba-Surma, M. Al-Mallah, and B. Dawn (2007) Adult Bone Marrow-Derived Cells for Cardiac Repair. A Systematic Review and Meta-analysis. *Archives of Internal Medicine* 167, 989–97.

Anon. (2002) Surgical Treatment for Parkinson's Disease: Neural Transplantation. *Movement Disorders* 17, S148–S155.

A. Atala (2006) Recent developments in tissue engineering and regenerative medicine. *Current Opinion in Pediatrics* 18, 167–71.

C. Bordignon (2006) Stem-cell therapies for blood diseases. *Nature* 441, 1100–2.

J. Couzin-Frankel (2010) Replacing an Immune System Gone Haywire. *Science* 327, 772–4.

C. E. P. Goldring, P. A. Duffy, N. Benvenisty, P. W. Andrews, U. Ben-David, R. Eakins, N. French, N. A. Hanley, L. Kelly, N. R. Kitteringham, J. Kurth, D. Ladenheim, H. Laverty, J. McBlane, G. Narayanan, S. Patel, J. Reinhardt, A. Rossi, M. Sharpe, and B. K. Park (2011) Assessing the Safety of Stem Cell Therapeutics. *Cell Stem Cell* 8, 618–28.

H. Green (2008) The birth of therapy with cultured cells. *BioEssays* 30, 897–903.

D. M. Harlan, N. S. Kenyon, O. Korsgren, B. O. Roep, and Immunology of Diabetes Society (2009) Current Advances and Travails in Islet Transplantation. *Diabetes* 58, 2175–84.

O. Lindvall and I. Hyun (2009) Medical Innovation Versus Stem Cell Tourism. *Science* 324, 1664–5.

O. Lindvall and Z. Kokaia (2006) Stem cells for the treatment of neurological disorders. *Nature* 441, 1094–6.

R. T. Maziarz and D. Driscoll (2011) Hematopoietic Stem Cell Transplantation and Implications for Cell Therapy Reimbursement. *Cell Stem Cell* 8, 609–12.

D. J. Mooney and H. Vandenburgh (2008) Cell Delivery Mechanisms for Tissue Repair. *Cell Stem Cell* 2, 205–13.

R. Passier, L. W. van Laake, and C. L. Mummery (2008) Stem-cell-based therapy and lessons from the heart. *Nature* 453, 322–9.

P. Rama, S. Matuska, G. Paganoni, A. Spinelli, M. D. Luca, and G. Pellegrini (2010) Limbal Stem-Cell Therapy and Long-Term Corneal Regeneration. *New England Journal of Medicine* 363, 147–55.

M. Ronaghi, S. Erceg, V. Moreno-Manzano, and M. Stojkovic (2010) Challenges of Stem Cell Therapy for Spinal Cord Injury: Human Embryonic Stem Cells, Endogenous Neural Stem Cells, or Induced Pluripotent Stem Cells? *Stem Cells* 28, 93–9.

Index

Stem Cells